AF411466

Life Cycle Assessment of Energy Systems

Life Cycle Assessment of Energy Systems

Editors

Guillermo San Miguel
Sergio Alvarez

MDPI • Basel • Beijing • Wuhan • Barcelona • Belgrade • Manchester • Tokyo • Cluj • Tianjin

Editors
Guillermo San Miguel
Universidad Politécnica of
Madrid
Spain

Sergio Alvarez
Universidad Politécnica of
Madrid
Spain

Editorial Office
MDPI
St. Alban-Anlage 66
4052 Basel, Switzerland

This is a reprint of articles from the Special Issue published online in the open access journal *Energies* (ISSN 1996-1073) (available at: https://www.mdpi.com/journal/energies/special_issues/ LCA_Energy_Systems).

For citation purposes, cite each article independently as indicated on the article page online and as indicated below:

LastName, A.A.; LastName, B.B.; LastName, C.C. Article Title. *Journal Name* **Year**, *Volume Number*, Page Range.

ISBN 978-3-0365-0524-4 (Hbk)
ISBN 978-3-0365-0525-1 (PDF)

Contents

About the Editors

Guillermo San Miguel is Lecturer and Senior Research Fellow (PCD-I3) at the School of Industrial Engineering (ETSII), Universidad Politécnica de Madrid. He holds a B.Sc. in Chemistry, an M.Sc. in Environmental Impact Assessment from University of Wales, and a Ph.D. in Environmental Engineering from Imperial College London. He was recipient of the Ramón & Cajal fellowship in 2003 and the I3 Award for Research Excellence in 2007 from the Spanish Ministry of Science. His research interests include Life Cycle Assessment (LCA); environmental, economic, and social analysis of products, services, and organizations; carbon footprint analysis; renewable energies; and waste management. In the last decade, he has coordinated numerous publicly and privately funded research projects and the Marie Curie network on sustainable energy. He has or is participating in and coordinating numerous technical organizations (e.g., esLCA) and international conferences (CEST2021, Global NEST, etc.). His work has led to the production of over 50 indexed articles, 90 conference papers, and 8 book/book chapters.

Sergio Alvarez is Assistant Professor at the Land Morphology and Engineering Department in the School of Civil Engineering, Universidad Politécnica de Madrid (UPM). He has an International PhD in Forest Engineering with distinction and the Extraordinary Doctorate Award. He has wide experience in sustainable studies under Life Cycle Assessment (LCA) and Multi-Regional Input–Output Analysis (MRIO Analysis). At present, he is a member of the Input–Output Analysis Society and Carbon Footprint UPM research group. He has participated in more than ten privately and public funded projects related to LCA and sustainability assessment covering a wide range of products and services (wood pallets, wood parquet, wildfire fighting, wind power, hydroelectric power, household consumption, civil infrastructures, and services). He is author of over 20 indexed articles and 6 books and book chapters. More specific info: www.huellaambiental.es

Preface to "Life Cycle Assessment of Energy Systems"

There is little doubt that the existing energy model, based on the mass consumption of fossil fuels, is utterly unsustainable. The urge for its profound transformation has intensified in recent years due to mounting evidence of global environmental degradation, potential shortages due to political instability in fossil fuel producing countries, and the economic consequences of higher prices due to a declining supply capacity. Despite unceasing warning signs, current projections from the International Energy Agency still describe a 1.3% yearly rise in energy demand until 2040, with fossil fuels remaining as the dominant source and expecting to account for 80% of the total primary energy supply in 2035. The result of a such trend will inevitably be a departure from the objectives of the 2016 UN Paris Agreement and an escalation in the strains exerted on the limits of our environment and our capacity to survive as a species.

In this context, several international initiatives are striving to redirect this situation so that a more sensible, beneficial future exists for all. For instance, the UN 2030 Agenda for Sustainable Development emphasizes, in Goal 7, the need to ensure universal access to affordable, reliable, and modern energy services. This document also states the need to increase the share of renewable energy and to improve efficiency, with actions required throughout the entire value chain of energy systems (including extraction of resources, transformation, transmission/transport, storage, and use). For all this to work, we need to develop advanced technologies and implement effective policy measures.

But, to ensure success, what will these new technologies and policies look? How can we ensure that the new technologies and plans are not flawed, that there is no transfer between impact categories and that the resulting scenario is more sustainable than the one we leave behind? How can we design the most sustainable technologies? How can they be deployed to maximize social wellbeing? How many jobs will be gained or lost in this energy transition? Will the economic cost compensate the environmental and social benefits? For this transition to be effective, all these questions and all the decisions that lay ahead cannot be taken lightly, and need to be responded to from a scientific, objective, and holistic perspective.

Life Cycle Thinking is a comprehensive and systemic framework that goes beyond the traditional focus on production sites and manufacturing processes to evaluate the sustainability of products and services. This framework has shaped a range of tools that are certainly applicable to investigating these questions and shedding light onto the sustainability assessment of energy systems. The most mature of these tools is the conventional Environmental Life Cycle Assessment (LCA), a robust procedure that is widely accepted and aimed at evaluating the attributional performance of systems, from the very simple to the highly complex. Even though it was born as a product-oriented tool focused solely on environmental issues, recent methodological extensions (such as Environmentally Extended Input–Output analysis, Hybrid IO–LCA, Consequential Analysis, Social LCA, and Environmental Life Cycle Costing) have broadened its scope and functionality.

This Special Issue on "LCA of Energy Systems" contains inspiring contributions describing the sustainability assessment of novel energy systems that are destined to shape the future energy system. These include battery-based and plug-in hybrid electric vehicles, geothermal energy, hydropower, biomass gasification, national electricity systems, and waste incineration. The identification and analysis of trends and singularities that result from these investigations will be invaluable to product designers, engineers, and policy makers. Furthermore, these exercises also contribute to refining the life cycle framework and harmonizing the methodological decisions that are specifically applicable to energy systems. We shall finish by sharing our hopes and desires that this analysis will promote the use of science and knowledge to shape a better world for everyone.

Guillermo San Miguel, Sergio Alvarez
Editors

Article

Comparative Life Cycle Energy and GHG Emission Analysis for BEVs and PhEVs: A Case Study in China

Siqin Xiong [1,2], Junping Ji [1,2,3,*] and Xiaoming Ma [1,2]

[1] School of Environment and Energy, Peking University Shenzhen Graduate School, Shenzhen 518055, China; xiongsiqin@pku.edu.cn (S.X.); xmma@pku.edu.cn (X.M.)
[2] College of Environmental Sciences and Engineering, Peking University, Beijing 100871, China
[3] Energy Analysis and Environmental Impacts Division, Lawrence Berkeley National Laboratory, One Cyclotron Road, MS90R2121, Berkeley, CA 94720, USA
*   Correspondence: jackyji@pku.edu.cn

Received: 2 January 2019; Accepted: 26 February 2019; Published: 3 March 2019

Abstract: Battery electric vehicles (BEVs) and plug-in hybrid electric vehicles (PHEVs) are seen as the most promising alternatives to internal combustion vehicles, as a means to reduce the energy consumption and greenhouse gas (GHG) emissions in the transportation sector. To provide the basis for preferable decisions among these vehicle technologies, an environmental benefit evaluation should be conducted. Lithium iron phosphate (LFP) and lithium nickel manganese cobalt oxide (NMC) are two most often applied batteries to power these vehicles. Given this context, this study aims to compare life cycle energy consumption and GHG emissions of BEVs and PHEVs, both of which are powered by LFP and NMC batteries. Furthermore, sensitivity analyses are conducted, concerning electricity generation mix, lifetime mileage, utility factor, and battery recycling. BEVs are found to be less emission-intensive than PHEVs given the existing and near-future electricity generation mix in China, and the energy consumption and GHG emissions of a BEV are about 3.04% (NMC) to 9.57% (LFP) and 15.95% (NMC) to 26.32% (LFP) lower, respectively, than those of a PHEV.

Keywords: life cycle assessment; battery electric vehicle (BEV); plug-in electric vehicle; energy; greenhouse gas (GHG) emissions

---

1. Introduction

Currently, China is the world's largest vehicle producer and sales market. However, the rapid growth of car ownership in recent years has raised grave concerns about national energy security, traffic safety, and climate change. According to statistics, China's reliance on oil importation exceeded 65 percent by the end of 2017 [1]. At the same time, the transport sector contributes to a significant share of the country's total greenhouse gas (GHG) emissions. Recently, the Chinese government has regarded electric vehicles (EVs) as the alternative to internal combustion engine vehicles (ICEVs) to diminish GHG emissions and to alleviate the dependence on gasoline. Since 2015, China has already become the largest EV market globally and the accumulated number of EVs exceeded 1 million at the end of 2017. Besides, in the energy saving and new energy automotive industry development plan 2012–2020 [2], it is estimated that the total production and sales of pure battery electric vehicles (BEVs) and plug-in electric vehicles (PHEVs) will amount to 5 million vehicles by 2020, 5 times more than the current ownership.

BEVs and PHEVs are two main types of EVs and are already commercially available. Noticeably, hybrid electric vehicles are seen as an extended model of ICEVs because they do not take electricity from the grid [3]. The choice of vehicle technologies depends on multi-aspect factors, including affordability, engineering performance, policy guidance, and environmental benefits. The differences surrounding the economic viability and electrochemistry performance of BEVs and PHEVs are clearly recognized. For example, the higher purchase cost is required for BEVs, relative to comparable PHEVs, but this additional cost can currently be compensated by higher subsidies and lower fuel costs in operation. On the other hand, the limited range of BEVs is a major challenge for the wide diffusion of BEVs. However, from the perspective of life cycle environmental performance analysis of BEVs and PHEVs, a consensus has not reached concerning which option has more energy saving and lower emissions.

Additionally, the supportive policies in current China give priority to BEVs, enhancing BEVs attractiveness for potential customers. In the early stage of deploying EVs, such government support played a determinant role to sway automakers to adjust the production strategies. Thereby, if the targets of energy conservation and emission reduction in the transportation sector are desired to be fulfilled by promoting the development of EVs, the identification of which powertrain option has larger energy and emission reduction potential is necessary.

A broad body of literature compares the energy consumption and environmental impact of BEVs, PHEVs with ICEVs in a life cycle perspective [3–6]. However, direct and detailed comparisons between BEVs and PHEVs are hardly observed. Secondly, the majority of relevant studies compare the BEVs and PHEVs by only considering the fuel cycle but disregard the vehicle cycle [7–9]. For example, Ke et al. (2017) [10] conducted a detailed Well-to wheels (WTW) analysis based on real-world data and found that Beijing's BEVs can significantly reduce WTW carbon dioxide emissions compared with their conventional gasoline counterparts, even in a coal-rich region. Among these papers regarding the fuel cycle, most conclusions demonstrate that BEVs are superior to PHEVs in terms of environmental performance, but if the vehicle cycle is counted, the findings may not be warranted since a larger battery is necessary to be produced for BEVs than a class-equivalent PHEV to overcome the range limitation. Thirdly, the preceding research regarding the fuel cycle of BEVs and PHEVs was almost based on European or U.S. cases and indicates that the results depend on the electricity profile and driving conditions of each specific case. For example, Onat et al. (2015) [11] compared various vehicle options across 50 states and concluded that EVs are the least carbon-intensive option in 24 states. Casals et al. (2016) [12] calculated the EV global warming potential for different European countries under various driving conditions and concluded that the current electricity profile in some countries (e.g., France or Norway) is well suited to accommodate EV market penetration, while countries like Germany and the Netherlands do not offer immediate GHG emission reductions for the uptake of EVs. In this sense, the advantages of BEVs may not be guaranteed in China, where the electricity mix is dominated by coal. As the most crucial part of EVs, the traction battery determines the environmental and engineering performance of vehicles. In the current Chinese traction battery market, lithium iron phosphate (LFP) and lithium nickel manganese cobalt oxide (NMC) are the two dominant battery chemistries, but these two battery types have different energy requirements in their production process, along with their unique electrochemistry features, which affect the energy demand of vehicles in the use stage. Therefore, specifically considering the battery chemistries is an important part of life cycle analysis of electric vehicles.

With the above information in mind, this study aims to comprehensively compare the life cycle energy consumption and GHG emission performance of BEVs and PHEVs, where both the fuel cycle and the vehicle material cycle are involved and two mainstream battery chemistries (LFP and NMC) are considered. Here, we attempt to address two questions:

Which electric vehicle technology corresponds to lower energy consumption and GHG emissions?

Will the relative outperformance of such vehicle technology change with the variation in battery chemistries, electricity mix, driving distance, and some other important factors?

2. Materials and Methods

Life cycle assessment (LCA) is a method to assess the life cycle potential environmental performance of a product or a service [13]. The standardized methodology defines four steps, the definition of the goal and scope, the life cycle inventory, the life cycle impact assessment and the interpretation of results. In this study, a comparison between BEVs and PHEVs is discussed by using the LCA approach to help us identify the superiority of these vehicle technologies in terms of energy savings and GHG emission reductions.

2.1. Goal and Scope

In this study, four electric vehicle types representing different vehicle technologies (BEV and PHEV) and battery options (LFP and NMC) have been discussed. Qin 300 (BEV-LFP), Qin 80 (PHEV-LFP), Qin 450 (BEV-NMC), and Qin 100 (PHEV-NMC) were chosen as the representative vehicles and the related information is mainly provided by its manufacturer, BYD, a major leading electric vehicle maker in China [14]. The choice of Qin series is due to its high market share, which contributed to 7% of the total new electric vehicles in the first half year of 2018. Especially in the PHEV market, Qin PHEV models account for 23.8% in the same period. Besides, choosing the vehicles from one plant allows a comparable basis for comparison, such as the comparative size and class of vehicles, the same modeling approach of energy efficiency, and unwanted variations in the production line are greatly avoided.

2.2. System Boundary

The system boundary includes both the fuel cycle and the vehicle cycle. The functional unit is expressed as per driven distance (per kilometers; per km) and GHG emissions are reported in grams CO_2 equivalents (g CO_2-eq).

Fuel life cycle

- Well to pump stage (WTT): The extraction, production and transport of feedstock, and the refining, production and distribution of gasoline and electricity
- Pump to wheels stage (TTW): The fuel utilized by vehicles in the use phase

Vehicle life cycle

- The production of raw materials
- The manufacturing of vehicle components, including the vehicle body, traction battery and fluids
- The assembly stage
- The distribution and transportation stage
- The maintenance of the vehicle throughout its life time
- The disposal of the vehicle, also known as the end-of-life stage

2.3. Life Cycle Inventory

2.3.1. The Fuel Cycle

The fuel cycle consists of the well-to-tank (WTT) stage and the tank-to-wheel (TTW) stage. As for the WTT stage, the primary energy including coal, liquefied gasoline gas, and natural gas are inputted to produce the terminal energy of gasoline and electricity. In 2017, the electricity mix in China is shown in Figure 1. The conversion efficiency of primary energy, the proportion of fuel consumption in various processes and the transportation distance of primary energy can be obtained or calculated based on the data from official yearbooks and other related publications [8,15,16].

As for the TTW stage, the fuel efficiencies of BEVs and PHEVs, as shown in Table 1, are provided by the car marker and have been verified through a fuel consumption record website, where the real-world energy efficiency data are reported by vehicle users [17]. The energy consumption and GHG emissions in the TTW stage are calculated by Equation (1).

$$E_{\text{TTW}} = E_{\text{electricity}} \times \text{UF} + \left(E_{\text{upstream}} + E_{\text{combustion}} \right) \times (1 - \text{UF}) \tag{1}$$

where E_{TTW} denotes the energy consumed per kilometer in the TTW stage, $E_{electricity}$ represents the upstream energy consumption of electricity, while $E_{upstream}$ and $E_{combustion}$ represent the upstream and the combustion emissions of gasoline, respectively. The utilization factor (UF) is defined as the distance fraction that is powered by electricity whereas (1-UF) represents the fraction of travel powered by gasoline [18]. For BEVs, the UF equals to 1 while that for PHEVs is assumed as 40% in this paper based on the assumption by Hou, Wang and Ouyang (2013) [18]. Similar methodology is applied to calculate GHG emissions. The Greenhouse Gases, Regulated Emissions, and Energy Use in Transportation Model (GREET) [19] is used to calculate the energy consumption and GHG emissions in the fuel cycle.

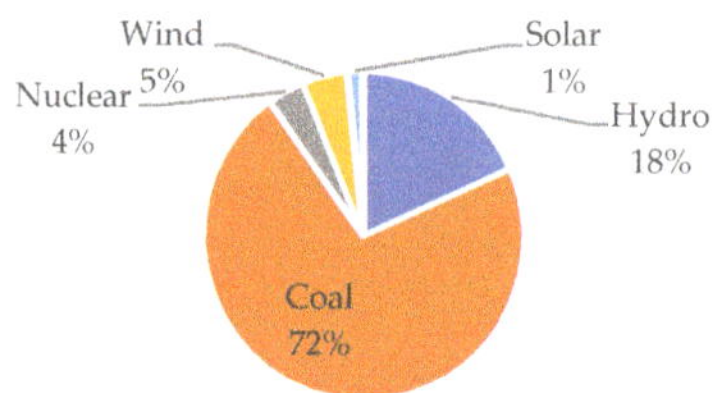

Figure 1. The electricity profile of China in 2017.

Table 1. The fuel efficiency of each vehicle technology.

Fuel Efficiency	Qin 300 (BEV)	Qin 80 (PHEV)	Qin 450 (BEV)	Qin 100 (PHEV)
Fuel efficiency (electricity) (kWh/100 km)	15.3	18.39	15.0	16.8
Fuel efficiency (gasoline) (L/100 km)	-	5.88	-	6.01

2.3.2. The Vehicle Cycle

As shown in Table 2, the vehicle and battery related parameters are provided by the car manufacturer, or assumed after personal communication with the car marker and car users.

Table 2. The key parameters of representative vehicles.

Parameters	Qin 300 (BEV)	Qin 80 (PHEV)	Qin 450 (BEV)	Qin 100 (PHEV)
Battery type	LFP	LFP	NMC	NMC
Total weight (kilogram, kg)	1950	1760	1950	1785
Battery weight (kg)	494	177	444	183
Battery capacity (kWh)	47.5	15.2	60.5	17.1
Capacity density (Wh/kg)	92.6	85.9	140.7	93.4
All-electric range (km)	300	80	400	100
Lifetime mileage (km) [1]	160,000	160,000	120,000	120,000

[1] The lifetime mileage is assumed by the author after personal communication with the car marker and some vehicle owners.

The vehicle cycle includes five phases: material production and the vehicle production, the vehicle assembly, the transportation of the vehicle, the maintenance service and the end-of-life stage.

As for the material production and the vehicle production stage (the vehicle production, for short), the inventory of pre-manufacturing, such as the raw material extracting and processing, is based on published studies and reports [9,19–22], and GaBi software [23], which is an LCA computational platform and accommodates thousands of background processes and elementary flows. This paper splits the vehicle into three parts: the vehicle body (excluding the battery and fluids), the battery and the fluids (including engine oil, brake fluid, transmission fluid, powertrain coolant, and wiper fluid); production-related inputs and outputs of each part are specified. Table 3 contains the list of materials for each vehicle technologies, and the material breakdown of vehicle body and fluids is based on the reports given by Sullivan and Gaines (2010) [24] Mayyas, et al. (2012) [25] while that of battery packs is based on estimations given by Peters, Baumann, Zimmermann, Braun and Weil (2017) [20], Peters and Weil (2018) [21], Majeau-Bettez, Hawkins and Str Mman (2011) [22]. It is noted that the main composition difference between BEVs and PHEVs is the powertrain, where a PHEV consists of both an electric motor and internal combustion engine, while a BEV is exclusively propelled by the electric motor. In the manufacturing phase, main material transformation processes of the vehicle body are considered, including the stamping, casting, forging, extrusion, and machining, and the inventory is estimated on the basis of previous reportedly data [24–27]. In terms of the battery packs, extensive studies have focused on the cell manufacturing and pack assembly stage. Among these studies, the modelling approach of energy demand (one is to allocate the total energy demand of a plant by its output; another is to use data from theoretical considerations for specific processes) is identified as a major cause of deviated results [20]. However, this comparative analysis will not be affected much by the modelling approach when these vehicles come from the same manufacturing plant. Therefore, we estimate the values based on an LCA review study reported by Peters and Weil (2018) [21]. By following these steps, the energy and GHG emissions associated with vehicle production stage are calculated by using GaBi software.

The assembly stage mainly includes stamping, welding, final assembly, injection molding, and painting. The production of heating, ventilation and air conditioning are not included in the comparative study since almost the same products are used for these different vehicles. In the assembly process, the energy consumption and GHG emissions are based on Mayyas, Omar, Hayajneh and Mayyas (2017) [25], J. L. Sullivan (2010) [28], Papasavva et al. (2002) [29].

The transportation of the vehicle includes two parts, from the production plant to the service shop, and from the maintenance shop to the dismantling sites [21]. The distance is set as 1600 km and 500 km, respectively, and diesel is assumed to be used in the road transportation.

Concerning the maintenance and replacement, we make assumptions based on previous studies, our communication with vehicle users and field investigation in the automobile service factory. As shown in Table 4, it is assumed that the tires and the engine oil should be replaced every 62,500 km, 6250 km, respectively and the wiper fluid, brake fluid, and powertrain coolant are completely consumed every 12,500 km, 62,500 km, and 62,500 km, respectively. In this paper, it is assumed that only one transmission oil is replaced during the life cycle of the car and the lifetime of the battery equals the lifetime of the vehicle.

For the end-of-life stage, this paper considers the energy consumption in the disassembly process and the avoided energy by recycling steel, aluminum, copper, and iron. Although batteries contain some valuable metals that need to be recycled, huge uncertainties exist when recycling activities are not conducted at a large scale. Additionally, most studies conclude that the end of life phase makes a small contribution to the whole life cycle [30–32]; therefore, we disregard the battery recycling in the baseline scenario but discuss it in the following sensitivity analysis. Besides, it is assumed that fluids, glasses and other non-metal materials are not recycled for their relatively cheap price. The energy consumption and regeneration rates are shown in Table 5, which are based on the recycling inventory reported by De Kleine et al. (2014) [33], Ruan et al. (2010) [34].

Table 3. The material component and the mass percentage of each vehicle technology (Unit: kg).

Vehicle Part	Component	Qin 300	Qin 80	Qin 450	Qin 100	Battery	Component	Qin 300	Qin 80	Qin 450	Qin 100
The vehicle body	Steel	943.41	1021.48	976.61	1034.08	Cathode	Positive active material	97.63	34.98	82.12	33.85
	Cast Iron	28.42	81.66	29.42	82.66		Carbon black	5.61	2.01	4.72	1.94
	Aluminium	92.35	100.15	95.6	101.38		Polytetrafluoroethylene (PTEF)	8.98	3.21	7.55	3.12
	Copper	66.78	66.25	69.13	67.07		N- Methyl pyrrolidone (NMP)	31.42	11.26	26.43	10.89
	Glass	49.73	46.22	51.48	46.79		Aluminium	16.28	5.83	14.64	6.04
	Plastic	171.92	163.31	177.97	165.33	Anode	Graphite	34.38	12.32	36.33	14.97
	Rubber	25.57	26.19	26.47	26.51		PTEF	1.81	0.65	1.92	0.79
	Others	42.62	29.27	44.12	29.63		NMP	10.14	3.63	10.71	4.42
	In total	1420.8	1534.53	1470.8	1553.45		Copper	37.55	13.45	33.77	13.92
Fluids	Engine oil	0	3.9	0	3.9	Electrolyte	Lithium Hexafluorophosphate	6.51	2.34	5.86	2.42
	Brake fluid	0.9	0.9	0.9	0.9		Ethylene carbonate	23.89	8.56	21.48	8.85
	Transmission fluid	0.8	0.8	0.8	0.8		Dimethyl carbonate	23.89	8.56	21.48	8.85
	Powertrain coolant	7.2	10.4	7.2	10.4	Separator	Polyethylene	7.46	2.67	6.72	2.77
	Wiper fluid	2.7	2.7	2.7	2.7		Polypropylene	7.46	2.67	6.72	2.77
	Additions	13.6	13.6	13.6	13.6	Shell	Polypropylene	20.93	7.50	18.92	7.80
	In total	25.2	32.3	25.2	32.3		Aluminium	146.48	52.48	132.43	54.58
						BMS	Copper	6.79	2.44	6.10	2.52
							Steel	5.43	1.94	4.88	2.02
							Circuit board	1.36	0.49	1.22	0.50
	In total	1446	1566.83	1496	1585.75		In total (Battery)	494	177	444	183
Total		1950	1760	1950	1785						

Table 4. The maintenance and replacement of vehicle materials.

Component	Qin 300	Qin 80	Qin 450	Qin 100
Tires	3	3	2	2
Engine oil	26	26	20	20
Wiper fluid	13	13	10	10
Brake fluid	3	3	2	2
Powertrain coolant	3	3	2	2
Gearbox	1	1	1	1
Battery	×	×	×	×

Table 5. The energy consumption in the end-of-life stage.

Energy Consumption	Steel	Aluminum	Copper	Iron
Coal (kg/kg)	-	-	-	-
Diesel fuel (kg/kg)	-	0.000031	-	-
Petrol (kg/kg)	-	0.000049	-	-
Natural gas (m^3/kg)	0.0066	0.0047	-	-
Electricity (kWh/kg)	1.18	0.22	2.65	0.62
Regeneration rate (%)	85.00	85.00	90.00	80.00

3. Results

In this section, the performances of different vehicle technologies are presented. The results of the fuel cycle are calculated in per km, while results of the vehicle cycle are firstly presented in the unit of per vehicle and then presented as per km by dividing the lifetime mileage of vehicles.

3.1. Fuel Cycle

Based on previous studies, the energy consumption and GHG emissions for BEVs and PHEVs in the fuel cycle are found to be primarily affected by the energy conversion efficiency, carbon intensity of fuels and the fuel efficiency of vehicles.

3.1.1. WTT Stage

In the WTT stage, terminal fuels are produced after primary energy acquisition and processing, transportation, power generation, transmission and distribution. As shown in Table 6, energy consumption and GHG emissions from gasoline and electricity production in China are calculated. The energy conversion efficiency of gasoline is about 88.4% while the calculated energy conversion efficiency of electricity is about 43.3%. Fossil energy consumption accounts for 86.31% of the total energy consumption in electricity production and is dominated by coal consumption. The calculated results are similar to previous studies [35]. Notably, the electricity used to power BEVs comes from a more energy and emission intensive source than gasoline in China. The production of 1 MJ electricity is 2.04 times higher energy demand than that of gasoline, along with 9.51 times more GHG emissions.

Table 6. The energy and emission intensities of gasoline and electricity production in 2017.

Fuel Type	Energy Intensity (MJ/MJ)	GHG Emissions Intensity (g CO_2-eq/MJ)
Gasoline	1.13	20.88
Electricity	2.31	198.65

3.1.2. TTW Stage

Based on the real-world fuel efficiency of EVs and the Equation (1), the energy required for BEV is 0.648 MJ/km and 0.635 MJ/km for LFP powered and NMC powered vehicles, respectively, whereas the corresponding energy consumption is 1.409 MJ/km and 1.406 MJ/km for PHEVs, that is about 2 times more energy is required for PHEVs to drive the same distance, relative to that of BEVs. In addition, PHEVs emit 113.92 g/km GHG emissions due to the use of gasoline in the TTW stage while no tailpipe emissions are exhausted in this stage for BEVs.

3.1.3. The Entire Fuel Cycle

In the overall perspective of the fuel cycle, the energy consumption and GHG emissions of PHEVs are higher than those of BEVs, as shown in Figure 2. The total energy consumption in the fuel cycle of BEV (LFP) is 1.50 MJ/km with 128.80 g/km GHG emissions, while that of PHEV (LFP) is 1.96 MJ/km with 190.58 g/km GHG emissions. The energy consumption of BEV (NMC) and PHEV (NMC) is 1.47 MJ/km and 1.92 MJ/km, along with 120.71 g/km and 185.86 g/km GHG emissions, respectively. It can be observed that BEVs have about 30% energy reduction benefits and about 50% GHG emission mitigation benefits relative to PHEVs in the fuel cycle.

As for the same vehicle technology coupled with different batteries, NMC-powered vehicles have more energy and emission reduction benefits compared with LFP-powered vehicles in the fuel cycle but the difference is negligible compared with the differences associated with the vehicle technology. It is worth noting that since the rank of batteries heavily relies on the assumed fuel efficiency, which is closely related to other vehicle characteristics; more information and detail analysis are required before the general conclusion is made.

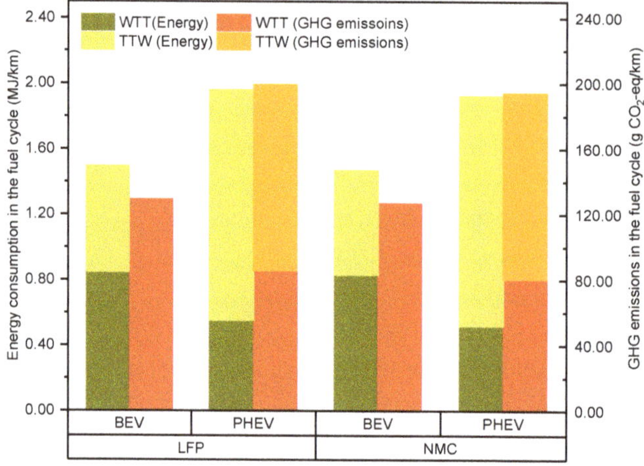

Figure 2. The energy consumption and GHG emissions in the fuel cycle.

3.2. Vehicle Cycle

3.2.1. Vehicle Body Production

Specifically, the vehicle body production accounts for a large proportion in terms of energy consumption and GHG emissions. The energy consumption of the vehicle body production is 57,600 MJ/vehicle, 60,000 MJ/vehicle, 62,400 MJ/vehicle and 62,400 MJ/vehicle for BEV (LFP), BEV(NMC), PHEV (LFP) and PHEV (NMC), respectively, and the corresponding proportion in the vehicle cycle is 35.13%, 33.48%, 48.59%, and 46.16%. Similarly, the GHG emissions from the vehicle body production are 3982 kg/vehicle, 4240 kg/vehicle, 4189 kg/vehicle and 4306 kg/vehicle, accounting for 33.99%, 31.35%, 47.17% and 44.08% for the above order of vehicles. Since PHEVs are heavier than the equivalent BEVs and the extra mass mainly comes from the internal combustion engine, the higher proportion of the vehicle body production for PHEVs could be attributed to the production of the internal combustion engine.

3.2.2. Battery Production

The energy consumption and GHG emissions from the battery production process also account for a large proportion of the vehicle cycle. The energy required to produce a battery is 50,920 MJ/vehicle, 67,566 MJ/vehicle, 18,245 MJ/vehicle and 27,848 MJ/vehicle, respectively. The associated GHG emissions of 3369 kg/vehicle, 5113 kg/vehicle, 1207 kg/vehicle and 2108 kg/vehicle, accounting for 28.76%, 38.26%, 13.43% and 21.58% of the total vehicle cycle. Due to the range limitation, heavier batteries are needed for BEVs than for PHEVs, and hence more energy is required to produce the battery, leading to more GHG emissions. For BEVs, the energy and emission contribution of the battery production are similar to those of the vehicle body production while the battery production for PHEVs contributes less than that for producing the vehicle body.

For the same vehicle technology with different battery chemistries, the energy consumption of NMC battery production is 152 MJ/kg coupled with 11.52 kg/kg GHG emissions, i.e., higher than that of an LFP battery with 103 MJ/kg energy consumption and 6.82 kg/kg GHG emissions. The difference is mainly because of the energy-intensive production process of the high cobalt-containing cathode of the NMC battery.

3.2.3. Fluids Production

The energy consumption and GHG emissions in the fluids production stage account for the smallest share of the vehicle cycle. About 1492.83 MJ/vehicle energy is consumed for BEVs compared with 1769.86 MJ/vehicle for PHEVs, along with 72.82 kg/vehicle and 91.43 kg/vehicle GHG emissions for BEVs and PHEVs, respectively; only about 1% of the energy and emissions contributes to the fluid production. Besides, PHEV consumes relatively more energy to produce fluids, mainly because of the additional needed for engine oil.

3.2.4. Assembly Stage

When it comes to the vehicle assembly stage, the energy consumption ranges from 20,376 MJ/vehicle to 22,301 MJ/vehicle for BEVs and PHEVs, with the GHG emission about 1800 kg/vehicle for BEVs and 1900 kg/vehicle for PHEVs. The higher energy requirement is associated with the heavier vehicle mass of PHEVs.

3.2.5. Transportation Stage

As for the transportation stage, 4077 MJ/vehicle energy is required for BEVs, along with 292 kg/vehicle GHG emissions while an average of 3706 MJ/vehicle energy is required for PHEVs, along with about 265 kg/vehicle GHG emissions. The transportation stage accounts for about 2.5% of the vehicle cycle energy consumption for all these four vehicle technologies.

3.2.6. Maintenance Stage

In the maintenance stage, 7640.03 MJ/vehicle and 5567.87 MJ/vehicle energy are needed for BEVs and 13,353.11 MJ/vehicle and 9942.62 MJ/vehicle for PHEVs; 505.72 kg/vehicle and 356.92 kg/vehicle are emitted from BEVs and 860.44 kg/vehicle and 628.05 kg/vehicle from PHEVs. PHEVs consume more energy than BEVs, since more fluids need to be supplied for PHEVs. Besides, LFP-powered vehicles need more replacement and consequently consume more energy than the NMC counterpart due to the longer lifetime mileage.

3.2.7. End of Life Stage

In the end-of-life stage, the energy required to dispose of the vehicles is counted, as well as the avoided energy by reusing some recycled metals in the production stage. The energy and emissions in the end-of-life stage are shown in Table 7.

Table 7. Energy consumption in the end-of-life stage.

Energy Consumption	BEV (LFP)	PHEV (LFP)	BEV (NMC)	PHEV (NMC)
Energy consumption (MJ/vehicle)	36,627.84	23,295.15	34,738.78	23,743.08
The avoided energy (MJ/vehicle)	−14,791.90	−16,346.10	−15,312.50	−16,547.70
Net value (MJ/vehicle)	21,835.93	6949.04	19,426.32	7195.39

3.2.8. Unit-Based Results in the Vehicle Cycle

In the vehicle cycle, the energy consumption of BEV (LFP), BEV (NMC), PHEV (LFP) and PHEV (NMC) is 163,941 MJ/vehicle, 179,199 MJ/vehicle, 128,433 MJ/vehicle and 135,188 MJ/vehicle, coupled with the GHG emissions of 11,712 kg/vehicle, 13,363 kg/vehicle, 8989 kg/vehicle and 9768 kg/vehicle, respectively.

As mentioned before, the lifetime mileage for LFP powered vehicles is about 160,000 km, while that for NMC powered vehicles is 120,000 km. As shown in Figure 3, LFP-powered BEVs consume 1.02 MJ/km, 27.65% higher than the LFP-powered PHEVs, with 0.80 MJ/km energy consumption. NMC-powered BEVs consume 1.49 MJ/km, 32.55% higher than NMC-powered PHEVs, whose energy consumption in the vehicle cycle is 1.13 MJ/km. The corresponding GHG emissions are 73.20 g/km, 56.18 g/km, 111.36 g/km and 81.40 g/km for BEV (LFP), BEV (NMC), PHEV (LFP) and PHEV (NMC).

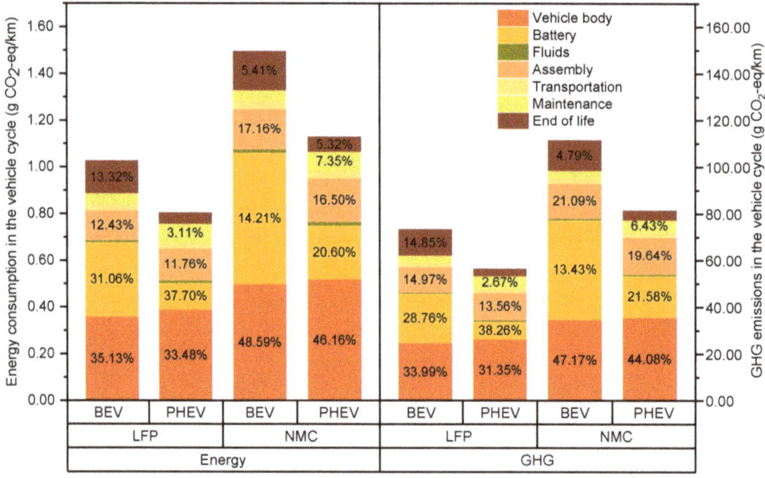

Figure 3. Energy consumption and GHG emissions in the vehicle cycle.

3.3. The Life Cycle

The entire life cycle energy consumption of BEV (LFP), BEV (NMC), PHEV (LFP) and PHEV (NMC) is 2.52 MJ/km, 2.96 MJ/km, 2.76 MJ/km and 3.05 MJ/km, respectively, along with 201.94 g/km, 237.57 g/km, 255.08 g/km and 275.46 g/km GHG emissions. About 20% GHG saving for LFP-powered BEVs is found compared with LFP-powered PHEVs and NMC-powered BEVs have 13.75% GHG emission reduction compared with their PHEV counterparts.

According to Table 8, where the contribution analysis of various processes is presented, the fuel cycle for BEVs has a similar share compared with the vehicle cycle in the life cycle energy consumption and a slightly higher proportion in terms of the life cycle GHG emissions; the fuel cycle is the dominant stage of the life cycle energy and emissions for PHEVs.

Table 8. Energy consumption and GHG emissions in the life cycle.

Vehicle Type	Fuel Cycle		Vehicle Cycle		Total	
Energy or GHG Emissions	Energy (MJ/km)	GHG (g CO₂-eq/km)	Energy (MJ/km)	GHG (g CO₂-eq/km)	Energy (MJ/km)	GHG (g CO₂-eq/km)
BEV (LFP)	1.50 (59.52%)	128.73 (63.75%)	1.02 (40.48%)	73.20 (36.25%)	2.52	201.93
PHEV (LFP)	1.96 (71.01%)	198.90 (77.98%)	0.80 (28.99%)	56.18 (22.02%)	2.76	255.08
BEV (NMC)	1.47(49.67%)	126.21 (53.13%)	1.49 (50.33%)	111.36 (46.87%)	2.96	237.57
PHEV (NMC)	1.92 (62.95%)	194.05 (70.45%)	1.13 (37.05%)	81.40 (29.55%)	3.05	275.46

4. Sensitivity Analyses

Sensitivity analyses are conducted to explore how the life cycle energy consumption and GHG emissions will be influenced by the uncertainty of key parameters, including the electricity mix, driving distance, and the recycling activities. In addition, break-even points between BEVs and PHEVs have been analysed.

4.1. Sensitivity Analysis of Electricity Profile

To achieve higher energy and emission reduction benefits, the shares of non-fossil power in the electricity mix should increase to a higher level. China has already introduced relevant policies and measures to develop a low-carbon electricity mix. For example, "The 13th five-year plan for electric power (2016–2020)" has proposed that the installed capacity of non-fossil fuel will be about 770 million kilowatts in 2020, accounting for about 39%, and the installed capacity of natural gas accounts for more than 5% while that of coal will decrease to about 55%. Based on relevant policies and previous studies, this study projects the electricity supply structure of China in the near future (2020) and in the long-term future (2030), as shown in Figure 4.

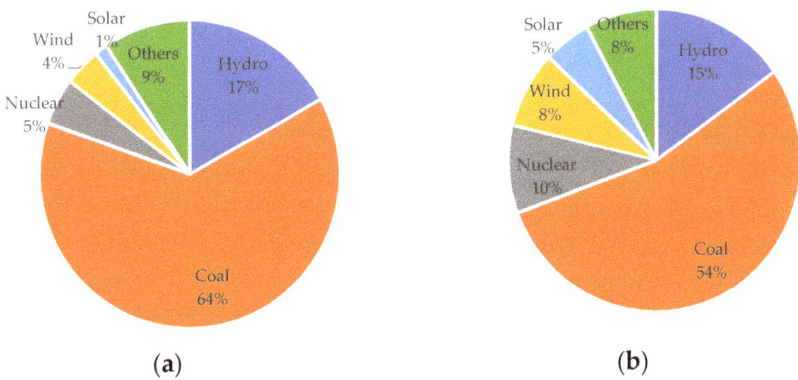

Figure 4. (**a**) The electricity profile of China in 2020; (**b**) The electricity profile of China in 2030.

Power generation technologies in 2020 are assumed to be at the same level as in 2017, but in 2030, advanced technologies and highly efficient equipment are expected to be implemented to improve the energy conversion efficiency and control the pollution emissions. Therefore, this study assumes the energy consumption and GHG emissions would be 10% less than the predicted baseline power generation mix in 2030. The energy and emission intensities of Chinese electricity in 2020 and 2030 are calculated by using the GREET model and the results are shown in Table 9. Therefore, the GHG emissions from power generation are 19.25% and 27.32% lower in 2020 and 2030 than those in 2017. Since the gasoline production technology is relatively mature, this study assumes that the gasoline production-related GHG emissions from in 2020 and 2030 will be the same as those in 2017.

Table 9. Energy and emission intensities of power generation in 2020 and 2030.

Energy or GHG Emissions	2017	2020	2030	2030 (Advanced Technologies)
Energy consumption (MJ/MJ)	2.31	2.25	2.12	1.91
GHG emissions (g/MJ)	198.65	182.34	160.41	144.37

As shown in Table 10, BEVs will generally achieve 6% and 9% emission reduction in 2020 and 2030, compared with 2017. The reduction benefits for PHEVs are lower, which are about 2.75% in 2020 and 3.80% in 2030. Clearly, the GHG emission differences between the BEVs and PHEVs expand to 31.40% (LFP) and 19.79% (NMC) in 2020. In 2030, the attractiveness of BEVs will be more prominent in that the emission reduction benefits of BEVs are expected to be 25.76% (NMC) – 40.00% (LFP) relative to PHEVs. In fact, if the electricity generation moves to a lower-emission intensity, the advantages of BEVs would be more remarkable.

Since China demonstrates a large amount of diversity in the electricity profiles, the conclusion may not be valid in all cities. Therefore, the break-even point is calculated to present in which cases BEVs outperform PHEVs in terms of the GHG emissions. Results obtained from break-even point analysis show that GHG emission intensity below 973.80 gCO_2-eq/kWh and 815.00 gCO_2-eq/kWh would make LFP- and NMC-powered BEVs, respectively, favorable options. According to Bauer et al. (2015) [36], where regional electricity profiles in China are analyzed, north, northeast, east, and northwest have about 900.00 gCO_2-eq/kWh GHG emission intensities in 2012 and cities like Beijing are estimated to have over 900.00 gCO_2-eq/kWh in 2020. Therefore, it is possible that PHEVs are currently preferable in parts of cities in China.

Table 10. The life cycle energy consumptions and GHG emissions in the 2020 and 2030 scenarios.

Year	2017		2020		2030	
Energy or GHG Emissions	Energy (MJ/km)	Emissions (g CO$_2$-eq/km)	Energy (MJ/km)	Emissions (g CO$_2$-eq/km)	Energy (MJ/km)	Emissions (g CO$_2$-eq/km)
BEV (LFP)	2.52	201.94	2.47	188.71	2.19	157.93
Change	-	-	−1.98%	−6.55%	−13.10%	−21.79%
PHEV (LFP)	2.76	255.08	2.73	247.94	2.35	226.96
Change	-	-	−1.09%	−2.80%	−14.86%	−11.02%
BEV (NMC)	2.96	237.57	2.91	223.75	2.61	191.61
Change	-	-	−1.69%	−5.82%	−11.82%	−19.35%
PHEV (NMC)	3.05	275.46	3.02	268.03	2.63	246.28
Change	-	-	−0.98%	−2.70%	−13.77%	−10.59%

4.2. Sensitivity Analysis of Driving Distance

The driving distance in this part includes two parts: the lifetime mileage and the all-electric ranges within one charging period. Because EVs have just come onto the market, real world data of lifetime mileage are unavailable. As stated in Table 2, the parameter of lifetime mileage is assumed and thus uncertainty is inevitable. Besides, PHEVs are able to use the battery in electric mode and consume gasoline when the battery charge is depleted [11]. Since the electric mode saves more energy with a lower fuel cost, drivers are often encouraged to use electricity as often as possible within the all-electric range. Therefore, the assumption of the travel distance for single travel and the all-electric range limitation are important parameters for the energy use and GHG emission rate of PHEVs.

4.2.1. Sensitivity Analysis of Lifetime Mileage

In the baseline scenario, the lifetime mileage is assumed to be a certain value and remains the same for BEVs and PHEVs provided that they use the same battery type. To deal with the uncertainty of the lifetime mileage, this parameter is considered as any value within an interval, and the range of life cycle GHG emissions of each vehicle is calculated accordingly. The equation of life cycle GHG emissions is shown as Equation (2).

$$GHG\ (life\ cycle)_i = \frac{GHG\ (vehicle\ cycle)_i}{r_i} + GHG\ (fuel\ cycle)_i \qquad (2)$$

where i relates to BEV-LFP, BEV-NMC, PHEV-LFP and PHEV-NMC; r_i represents the lifetime mileage of each vehicle, with the assumed range of [120,000 km, 160,000 km]; $GHG\ (life\ cycle)_i$, $GHG\ (vehicle\ cycle)_i$ and $GHG\ (fuel\ cycle)_i$ represent GHG emissions of each vehicle for the life cycle, vehicle cycle and fuel cycle, respectively.

The calculated ranges of life cycle GHG emissions for each vehicle are shown in Table 11. Since the minimum emissions of PHEVs exceed the maximum values of their counterpart BEVs, it is highly likely that BEVs outperform PHEVs from the life cycle perspective.

Table 11. The calculated range of life cycle GHG emissions for each vehicle.

GHG Emissions	BEV (LFP)	PHEV (LFP)	BEV (NMC)	PHEV (NMC)
Minimum (g CO$_2$-eq/km)	201.93	255.08	209.73	255.10
Maximum (g CO$_2$-eq/km)	226.33	273.81	237.57	275.45

4.2.2. Sensitivity Analysis of All-Electric Range

The all-electric ranges for one charging period for our representative vehicles are reported as 300 km, 400 km for BEVs and 80 km, 100 km for PHEVs. Under real-world driving conditions, 80% depth discharge is always applied. Therefore, we assume that 80% of the reported ranges are reached within each driven trip. The UF for PHEVs can be assumed as follows:

$$UF(R_{LFP}) = \begin{cases} 1, & 0 < R_{LFP} \leq 64; \\ 64/R_{LFP}, & R_{LFP} > 64 \end{cases} \qquad (3)$$

$$UF(R_{NMC}) = \begin{cases} 1, & 0 < R_{NMC} \leq 80; \\ 80/R_{NMC}, & R_{NMC} > 80 \end{cases} \qquad (4)$$

where $UF(R_{LFP})$ and $UF(R_{NMC})$ represent the UF of LFP-powered and NMC-powered PHEVs, respectively; R_{LFP} and R_{NMC} represent the travel distances for each time.

Since BEVs are only able to use electricity to propel the vehicles, the life cycle energy use and GHG emissions on the basis of per km will not change. Our results show that for LFP-powered vehicles, as long as the driven distance is below the range limitation of BEVs (300 km), the BEV has lower life cycle emissions than the PHEV. For NMC-powered vehicles, when the driven distance is below 80 km, the life cycle GHG emissions is 222.76 g CO$_2$-eq/km for the PHEV, 6.65% less than the life cycle emissions of the BEV. When the driven distance is higher than 80 km but less than 96.34 km, the per-km-based emissions result for the PHEV increases with the distance, but is still below that of the BEV. Therefore, the break-even point for UF(NMC) is 0.83, at the point that the travel distance reaches 96.23 km. Additionally, it is highlighted that PHEVs would be the option when the travel distances exceed the range limitation of BEVs.

4.3. Sensitivity Analysis of the Recycling Process

Although battery recycling activities are still in their infancy and huge uncertainties are related to the recycling techniques, it is widely believed that traction batteries are worthy of recycling and reusing, both from the environmental perspective and the cost-benefit view. Based on previous studies, the NMC batteries, which contain high cobalt and nickel, are expected to be recycled and about 50%

energy for the battery's primary production is reported to be saved. However, the LFP batteries are hardly reused since the lithium metal is relatively abundant and cheap [37–41].

In this sense, we explore whether the consideration of recycling of batteries would considerably change the results. Here, we assume 30% energy saving for NMC batteries in the battery production process and 10% for LFP, as part of lithium, nickel, and aluminum are also recyclable. Considering the battery recycling, the life cycle energy consumption and GHG emissions of four vehicle types are shown in Figure 5:

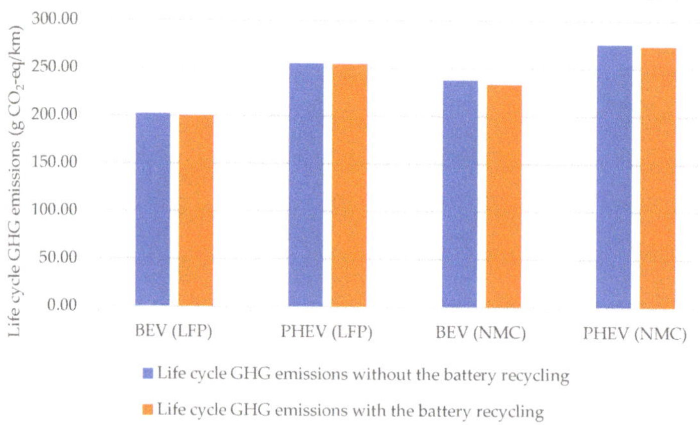

Figure 5. Life cycle GHG emissions with battery recycling.

Notably, battery recycling offers some environmental benefits to electric vehicles, but such a contribution accounts for little in the life cycle. Since heavier batteries are installed in BEVs, relative to PHEVs, more environmental benefits could be achieved through battery recycling. Therefore, the superiority of BEVs is further confirmed. Besides, although NMC-powered vehicles have greater emission reduction, they are still not able to exceed the LFP-powered ones in terms of GHG emission performance.

In general, sensitivity analyses have been performed concerning future electricity generation pathways (2020 and 2030), lifetime mileage, travel distance and UF, and expended system boundary of the recycling stage. The main conclusions can be summarized as follows:

(1) As long as the emission intensity of the power generation is less than 815.00 g CO_2-eq/kWh, BEVs are more competitive than PHEVs for both batteries in terms of GHG emissions.

(2) When the lifetime mileage is within 120,000 km to 160,000 km, which is reasonable for vehicles, BEVs emit less GHG emissions than PHEVs. In terms of the travel distance at each time, LFP-powered BEVs are superior to PHEVs, as long as the distance is below the range limitation, while NMC-powered PHEVs are better if the driven distance during single travel is under 96.23 km; at this point, the UF is 0.83.

(3) The impacts of battery recycling are found to be small from the life cycle perspective.

5. Conclusions

In this work, a comprehensive life cycle analysis is conducted to compare BEVs and PHEVs. This analysis is divided into two parts: fuel cycle and vehicle cycle, performed with two different battery chemistries cases: LFP and NMC, and framed to China. The main conclusions are drawn as follows:

(1) BEVs are currently better choices than PHEVs, in terms of energy consumption and GHG emissions. Specifically, BEVs have 3.04% (NMC) to 9.57% (LFP) energy mitigation benefits and 15.95% (NMC) to 26.32% (LFP) emission reduction benefits compared to PHEVs.

(2) The fuel cycle and vehicle cycle have similar contributions to the life cycle emissions for BEVs while the fuel cycle is the dominant emission stage for PHEVs.

(3) Through sensitivity analyses, the superiority of BEVs is further confirmed as BEVs have lower GHG emissions than PHEVs in the vast majority of cases. In this study, NMC-powered PHEVs might be preferable if the GHG emission intensity is higher than 815.00 g CO_2-eq/kWh, or when the driven distance at a single travel is over 96.23 km.

While this study provides a comprehensive life cycle environmental performance comparison, some limitations remain.

(1) Although the selected vehicles are believed to be representative, a larger number of vehicles should be considered to confirm the robustness of the results.

(2) Another source of variability in the results relates to battery lifetime assumptions. Since there is no practical evidence regarding the lifetime of batteries and the uncertainty relates to use patterns, future research should pay more attention to these aspects.

(3) Since GHG emission reduction is the main purpose of developing electric vehicles, other potential environmental impacts are disregarded in this study. If a more comprehensive comparison is desired, other impacts should be included.

Author Contributions: S.X. conceived and designed the work, analyzed the data and prepared the original draft; J.J. mentored the use of the LCA software and reviewed the writing; X.M. supervised all work. Conceptualization, S.X.; Data curation, S.X.; Formal analysis, S.X.; Methodology, J.J.; Software, J.J.; Supervision, J.J. and X.M.; Validation, X.M.; Writing—original draft, S.X.; Writing—review & editing, J.J.

Funding: This research received no external funding.

Acknowledgments: We wish to thank Master Ying Duan for helping us with the GaBi software.

References

1. Gov, E. China—Oil and Gas | export.gov. Available online: https://www.export.gov/article?id=China-Oil-and-Gas (accessed on 20 August 2018).
2. China, I. Energy Saving and New Energy Automotive Industry Development Plan 2012–2020. Available online: https://www.iea.org/policiesandmeasures/pams/china/name-32249-en.php (accessed on 20 August 2018).
3. Ma, H.; Balthasar, F.; Tait, N.; Riera-Palou, X.; Harrison, A. A new comparison between the life cycle greenhouse gas emissions of battery electric vehicles and internal combustion vehicles. *Energy Policy* **2012**, *44*, 160–173. [CrossRef]
4. Noshadravan, A.; Cheah, L.; Roth, R.; Freire, F.; Dias, L. Stochastic comparative assessment of life-cycle greenhouse gas emissions from conventional and electric vehicles. *Int. J. Life Cycle Assess.* **2015**, *20*, 854–864. [CrossRef]
5. Mamalis, C.I.C.K. Environmental and economic effects of widespread introduction of electric vehicles in Greece. *Eur. Transp. Res. Rev.* **2014**, *6*, 365–376.
6. Messagie, M.; Boureima, F.S.; Coosemans, T.; Macharis, C.; Mierlo, J.V. A Range-Based Vehicle Life Cycle Assessment Incorporating Variability in the Environmental Assessment of Different Vehicle Technologies and Fuels. *Energies* **2014**, *7*, 1467–1482. [CrossRef]
7. Hao, H.; Qiao, Q.; Liu, Z.; Zhao, F. Impact of recycling on energy consumption and greenhouse gas emissions from electric vehicle production: The China 2025 case. *Resour. Conserv. Recycl.* **2017**, *122*, 114–125. [CrossRef]
8. Peng, T.; Ou, X.; Yan, X. Development and application of an electric vehicles life-cycle energy consumption and greenhouse gas emissions analysis model. *Chem. Eng. Res. Des.* **2018**, *131*, 699–708. [CrossRef]
9. Qiao, Q.; Zhao, F.; Liu, Z.; Jiang, S.; Hao, H. Cradle-to-gate greenhouse gas emissions of battery electric and internal combustion engine vehicles in China. *Appl. Energy* **2017**, *204*, 1399–1411. [CrossRef]
10. Ke, W.; Zhang, S.; He, X.; Wu, Y.; Hao, J. Well-to-wheels energy consumption and emissions of electric vehicles: Mid-term implications from real-world features and air pollution control progress. *Appl. Energy* **2017**, *188*, 367–377. [CrossRef]

11. Onat, N.C.; Kucukvar, M.; Tatari, O. Conventional, hybrid, plug-in hybrid or electric vehicles? State-based comparative carbon and energy footprint analysis in the United States. *Appl. Energy* **2015**, *150*, 36–49. [CrossRef]

12. Casals, L.C.; Martinez-Laserna, E.; García, B.A.; Nieto, N. Sustainability analysis of the electric vehicle use in Europe for CO_2 emissions reduction. *J. Clean. Prod.* **2016**, *127*, 425–437. [CrossRef]

13. ISO. *ISO 14040: Environmental Management—Life Cycle Assessment—Requirements and Guidelines*; International Organization for Standardization: Geneva, Switzerland, 2006.

14. BYD BYD Europe | BYD Official Web Site. Available online: http://www.bydeurope.com/ (accessed on 26 August 2018).

15. Li, X.; Ou, X.; Zhang, X.; Zhang, Q.; Zhang, X. Life-cycle fossil energy consumption and greenhouse gas emission intensity of dominant secondary energy pathways of China in 2010. *Energy* **2013**, *50*, 15–23. [CrossRef]

16. Ou, X.; Yan, X.; Zhang, X. Life-cycle energy consumption and greenhouse gas emissions for electricity generation and supply in China. *Appl. Energy* **2011**, *88*, 289–297. [CrossRef]

17. Xiaoxiongyouhao. Fuel Consumption Calculator_Actual Fuel Consumption Data and Statistical Reports. Available online: https://www.xiaoxiongyouhao.com/ (accessed on 26 August 2018).

18. Hou, C.; Wang, H.; Ouyang, M. Survey of daily vehicle travel distance and impact factors in Beijing. *IFAC Proc. Vol.* **2013**, *46*, 35–40. [CrossRef]

19. Semmens, J.; Bras, B.; Guldberg, T. Vehicle manufacturing water use and consumption: An analysis based on data in automotive manufacturers' sustainability reports. *Int. J. Life Cycle Assess.* **2014**, *19*, 246–256. [CrossRef]

20. Peters, J.F.; Baumann, M.; Zimmermann, B.; Braun, J.; Weil, M. The environmental impact of Li-Ion batteries and the role of key parameters—A review. *Renew. Sustain. Energy Rev.* **2017**, *67*, 491–506. [CrossRef]

21. Peters, J.F.; Weil, M. Providing a common base for life cycle assessments of Li-Ion batteries. *J. Clean. Prod.* **2018**, *171*, 704–713. [CrossRef]

22. Majeau-Bettez, G.; Hawkins, T.R.; Str Mman, A.H. Life cycle environmental assessment of lithium-ion and nickel metal hydride batteries for plug-in hybrid and battery electric vehicles. *Environ. Sci. Technol.* **2011**, *45*, 4548–4554. [CrossRef] [PubMed]

23. GaBi. Life Cycle Assessment LCA Software: GaBi Software. Available online: http://www.gabi-software.com/america/index/ (accessed on 4 July 2018).

24. Sullivan, J.L.; Gaines, L. *A Review of Battery Life-Cycle Analysis: State of Knowledge and Critical Needs*; Argonne National Laboratory: Argonne, IL, USA, 2010.

25. Mayyas, A.; Qattawi, A.; Omar, M.; Shan, D. Design for sustainability in automotive industry: A comprehensive review. *Renew. Sustain. Energy Rev.* **2012**, *16*, 1845–1862. [CrossRef]

26. Rydh, C.J.; Sandén, B.A. Energy analysis of batteries in photovoltaic systems. Part I: Performance and energy requirements. *Energy Convers. Manag.* **2005**, *46*, 1957–1979. [CrossRef]

27. Mayyas, A.; Omar, M.; Hayajneh, M.; Mayyas, A.R. Vehicle's lightweight design vs. electrification from life cycle assessment perspective. *J. Clean. Prod.* **2017**, *167*, 687–701. [CrossRef]

28. Sullivan, J.L.; Burnham, A.; Wang, M. *Energy-Consumption and Carbon-Emission Analysis of Vehicle and Component Manufacturing*; Center for Transportation Research, Energy Systems Division, Argonne National Laboratory: Chicago, IL, USA, 2010.

29. Papasavva, S.; Kia, S.; Claya, J.; Gunther, R. Life cycle environmental assessment of paint processes. *J. Coat. Technol.* **2002**, *74*, 65–76. [CrossRef]

30. Hawkins, T.R.; Singh, B.; Majeau-Bettez, G.; Mman, A.S. Comparative Environmental Life Cycle Assessment of Conventional and Electric Vehicles. *J. Ind. Ecol.* **2012**, *17*, 53–64. [CrossRef]

31. Aguirre, K.; Eisenhardt, L.; Lim, C.; Nelson, B.; Norring, A.; Slowik, P.; Tu, N. *Life cycle Analysis Comparison of a Battery Electric Vehicle and a Conventional Gasoline Vehicle*; California Air Resource Board: Sacramento, CA, USA, 2012.

32. Amarakoon, S.; Smith, J.; Segal, B. *Application of Life-Cycle Assessment to Nanoscale Technology: Lithium-Ion Batteries for Electric Vehicles*; US Environmental Protection Agency: Washington, DC, USA, 2013.

33. De Kleine, R.D.; Keoleian, G.A.; Miller, S.A.; Burnham, A.; Sullivan, J.L. Impact of Updated Material Production Data in the GREET Life Cycle Model. *J. Ind. Ecol.* **2014**, *18*, 356–365. [CrossRef]

34. Ruan, R.; Zhong, S.; Wang, D. Life cycle assessment of copper extraction from biological and pyrometallurgical processes. *Multipurp. Util. Miner. Resour.* **2010**, *39*, 33–37.

35. Liu, F.; Zhao, F.; Liu, Z.; Hao, H. China's Electric Vehicle Deployment: Energy and Greenhouse Gas Emission Impacts. *Energies* **2018**, *11*, 3353. [CrossRef]

36. Bauer, C.; Hofer, J.; Althaus, H.; Del Duce, A.; Simons, A. The environmental performance of current and future passenger vehicles: Life cycle assessment based on a novel scenario analysis framework. *Appl. Energy* **2015**, *157*, 871–883. [CrossRef]

37. Dunn, J.B.; Gaines, L.; Sullivan, J.; Wang, M.Q. Impact of Recycling on Cradle-to-Gate Energy Consumption and Greenhouse Gas Emissions of Automotive Lithium-Ion Batteries. *Environ. Sci. Technol.* **2012**, *46*, 12704–12710. [CrossRef] [PubMed]

38. Dewulf, J.; Van der Vorst, G.; Denturck, K.; Van Langenhove, H.; Ghyoot, W.; Tytgat, J.; Vandeputte, K. Recycling rechargeable lithium ion batteries: Critical analysis of natural resource savings. *Resour. Conserv. Recycl.* **2010**, *54*, 229–234. [CrossRef]

39. Simon, B.; Weil, M. Analysis of materials and energy flows of different lithium ion traction batteries. *Revue de Métallurgie* **2013**, *110*, 65–76. [CrossRef]

40. Fisher, K.; Wallén, E.; Laenen, P.P.; Collins, M. *Battery Waste Management Life Cycle Assessment*; Environmental Resources Management (ERM): Oxford, UK, 2006.

41. Gaines, L.; Sullivan, J.; Burnham, A.; Belharouak, I. Life-Cycle Analysis for Lithium-Ion Battery Production and Recycling. In Proceedings of the Transportation Research Board 90th Annual Meeting, Washington, DC, USA, 23–27 January 2011.

Article

Thermodynamic-Based Exergy Analysis of Precious Metal Recovery out of Waste Printed Circuit Board through Black Copper Smelting Process

Maryam Ghodrat [1],*, Bijan Samali [1], Muhammad Akbar Rhamdhani [2] and Geoffrey Brooks [2]

[1] Centre for Infrastructure Engineering, School of Computing, Engineering and Mathematics, Western Sydney University, Sydney 2751, Australia; B.samali@westernsydney.edu.au
[2] Department of Mechanical Engineering and Product Design, Swinburne University of Technology, Victoria 3122, Australia; arhamdhani@swin.edu.au (M.A.R.); gbrooks@swin.edu.au (G.B.)
* Correspondence: m.ghodrat@westernsydney.edu.au

Received: 7 March 2019; Accepted: 30 March 2019; Published: 5 April 2019

Abstract: Exergy analysis is one of the useful decision-support tools in assessing the environmental impact related to waste emissions from fossil fuel. This paper proposes a thermodynamic-based design to estimate the exergy quantity and losses during the recycling of copper and other valuable metals out of electronic waste (e-waste) through a secondary copper recycling process. The losses related to recycling, as well as the quality losses linked to metal and oxide dust, can be used as an index of the resource loss and the effectiveness of the selected recycling route. Process-based results are presented for the emission exergy of the major equipment used, which are namely a reduction furnace, an oxidation furnace, and fire-refining, electrorefining, and precious metal-refining (PMR) processes for two scenarios (secondary copper recycling with 50% and 30% waste printed circuit boards in the feed). The results of the work reveal that increasing the percentage of waste printed circuit boards (PCBs) in the feed will lead to an increase in the exergy emission of CO_2. The variation of the exergy loss for all of the process units involved in the e-waste treatment process illustrated that the oxidation stage is the key contributor to exergy loss, followed by reduction and fire refining. The results also suggest that a fundamental variation of the emission refining through a secondary copper recycling process is necessary for e-waste treatment.

Keywords: thermodynamic modeling; exergy; e-waste; secondary copper smelting; precious metal recovery; printed circuit board

---

1. Introduction

According to Rosen and Dincer [1], "exergy is an ultimate extent of work that can be generated by a flow of heat or work when it reaches an equilibrium state with an environment chosen as a reference". The exergy value is able to disclose the possibility of designing more efficient processes as well as identifying the threshold by which we can achieve these designs. The design process mainly consists of identifying and decreasing the sources of inefficiency in existing systems. The most systematic approach as recommended by many researchers (e.g., Szargut et al. [2]; Edgerton [3]) is to relate the second thermodynamic law and the impact on the environment by means of an exergy analysis. Several researchers used exergy as a thermodynamic-based index to describe the environmental impact assessment [4–19]. In 1997, Rosen and Dincer indicated that the concept of exergy could be reflected as a gauge for measuring the possible environmental impact of waste emissions [1], and the same researchers further emphasized that exergy characterized in the emission of the waste specifies how far the emissions and the considered reference environment are from each other. This also signifies the possible environmental variation as stated by Rosen and Dincer [20].

Other researchers such as Ji et al. [18] carried out a comprehensive comparison between standard chemical exergy and the environmental pollutant cost for contaminants that affect the atmosphere. Based on their research, the emitted exergy to the environment is considered to be an unrestrained dynamic likelihood for environmental destruction. Daniel and Rosen [9] studied the emissions generated in the lifespans of 13 automobile fuels and mapped the significance of the waste emissions exergy, which signified their imbalance with the environment.

Considering the greenhouse gas emissions as an environmental impact indicator, Rosen et al. indicated that exergy incorporated in emissions has a relatively high effect on the availability of the net exergy in the ecological community that corresponded to earth solar radiation [7,8]. In the meantime, Ayres et al. [4] and Ayres et al. [21] recognized that exergy might be utilized to combine waste, and waste exergy is a delegation for their possible damage to the ecosystem. These researchers recommended that the ratio between the exergy embedded in the waste outputs and that contained in the input is the most relevant benchmark for quantifying pollution.

There have also been various assessments regarding exergy losses throughout recycling that suggest improving the resource efficiency of several production processes. For example, Amini et al. [22] carried out an exergy analysis to quantify the material quality loss and efficiency of the resources in some recycling streams. They demonstrated the influence of contaminations on the amount of exergy of recovered materials. A light passenger car was chosen, and the weights of the various materials dropped in a crusher, which then went through the recycling steps and to the landfill, were calculated. The results of their study demonstrated that the amount of chemical exergy drops during various recycling steps. Some other researchers such as Ignatenko et al. [23] evaluated the efficiency of recycling systems with the aid of exergy. The same authors, Castro et al. [24] and Ignatenko et al. [23], assessed a number of automobile recycling schemes using their proposed optimization methods for recycling. Their results illustrated the ability of exergy analysis to support the evaluation of recycling systems. Castro et al. [24] proposed a technique to measure the amount of exergy and exergy losses of metal solutions through the process of recovery and recycling. They showed that the losses coming from recycling can be utilized as a key to the material quality loss and the effectiveness of the recovery system. The copper smelting industry has unique features that make the pyrometallurgical process of electronic wastes feasible [25], as it is categorized by a high consumption of thermal energy largely due to the high temperatures needed to produce cathode copper (99% pure copper) and precious metals, which are the main purpose of the electronic waste (e-waste) treatment process. The use of electronic wastes, especially waste printed circuit boards (PCBs), in the copper smelting industry has a drawback. Their disadvantage relates to containing a big amount of plastic and polymer parts; burning them releases a considerable amount of off-gas to the environment. The current work offers an exergy-based inclusive analysis of the waste gas, metal, and oxide dusts emissions from non-renewable fuel e-waste and metal scrape consumption in a proposed e-waste treatment through a pyrometallurgical process, which is the black copper smelting or secondary copper recycling process. The exergy balance of the proposed metallurgical route has been calculated using the HSC Chemistry Sim 8.0 thermochemical package (HSC and HSC Sim 7.1&8). The element distribution in the different inflow was predicted by the equilibrium calculations implemented by using the Fact-Sage 6.4 thermochemical package [26]. Two metal recovery scenarios (secondary copper recycling with 50% and 30% waste PCBs) have been evaluated using the developed thermodynamic model.

2. Exergy Perception

Measurement of Exergy Losses in Recycling Process

According to research done by Amini et al. [22], throughout the recycling of metallurgic metals, resource depletion is specifically estimated by multiplying the amount of chemical exergy of the depleted materials by the mass of that material. It is well known that the losses that occurred during the smelting process are due to contaminants. These contaminants melt in the liquefied metal and escalate the alloy entropy (i.e., enhancing the system disturbance); therefore, this kind of material or resource depletion is called the quality loss of the process, which is matched to an exergy loss due to the growth of the chaos within the process. Based on work done by Amini et al. [22], the losses happening during the pyrometallurgical process can be estimated by Equation (1):

$$\Delta E_{chemical} = (E_{chemical})input - (E_{chemical})output \tag{1}$$

In that, *input* and *output* denote the state of the process before and after the pyrometallurgical process, respectively. There is another kind of the loss that is linked to the amount of metals that were depleted during slag generation.

The conception of exergetic efficiency is considered as the origin of exergy balance for the input and output streams, in which *I* is called the irreversibility [27]:

$$E_{input} = E_{output} + I \tag{2}$$

The effectiveness of exergy is the described as the percentage of the total output exergy to the total input exergy throughout the recovering process [27]:

$$\eta = E_{output} / E_{input} \tag{3}$$

Therefore, E_{output} is the amount of chemical exergy of the ultimate product, where E_{input} is estimated as the summation of the chemical and accumulative exergy of the materials coming into the process. In all real processes, the entropy enhances every time that an actual process happens. This phenomenon could be interpreted as a drop of the obtainable exergy in the process. The preferred condition is that the exergy depletion is as insignificant as possible after each stage of the process.

For assessing the exergy losses during recycling, the focus has been drawn to the metal and oxide dust, as these are the most relevant materials considered in metal recycling out of e-waste, from an environmental point of view. When metals recycling is carried out, various types of emissions occur. These emissions can be divided into off-gas, exhaust, and metal and oxide dust. The losses during the metal recovery through the pyrometallurgical route can be schematized in Figure 1.

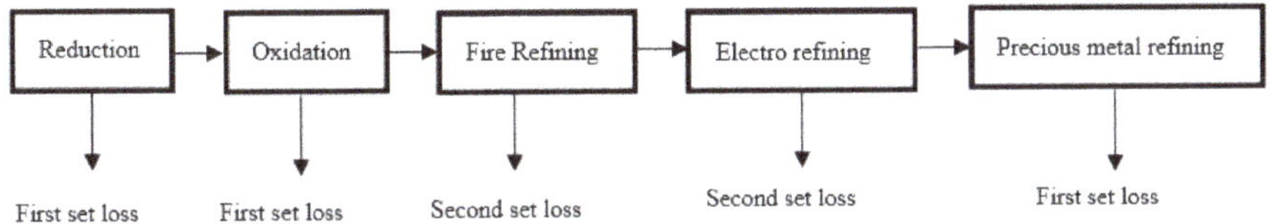

Figure 1. Exergy depletion through metal recycling out of electronic waste.

The first set of losses linked to the depleted materials that ended up in the landfill after the precious metal refining stage, and depleted in the slag through the pyrometallurgical process (reduction and oxidation stages). These losses lead to a decline in the amount of exergy in the process that is attributable to the material depletion from the process. The second set of losses arises during the refining stage (fire refining and electrorefining). When pollutants exist, they melt in the liquid metal that causes the disorder of the process to grow. Thus, the second sets of losses occur as the quality losses of the process and are correlated with an exergy loss caused by the rise of the disorder within the process. It is worthwhile to mention that this kind of loss is not related to the loss of mass. The procedure that has been considered in this study is based on the modeled process, and cannot be considered a common procedure. It is of interest to note that the exergies that are measured for a particular process-dependent model are applicable merely to that specific method. They cannot logically be correlated with exergies obtained from other process-dependent models. Finding the exergy of emitted wastes is the main focus of this paper; hence, the analysis employed here can detect the phases of the modeled process where the irreversibility or exergy destruction is the highest.

3. Process Description

To assess the influence of waste emission on the exergy content of recycled metals, an electronic waste processing plant set in a secondary copper smelting route has been considered in this study. The overall annual input materials of 110,000 t has been chosen and fed into the plant. Secondary copper is defined as the copper that has not originated from primary sources. So, copper from any sort of metallurgical, industrial, or consumer waste is considered secondary copper. Smelting in a shaft furnace is the most common pyrometallurgical route for refining copper from secondary sources [28–31]. However, the process has its own issues, such as impurities associated with the secondary materials. Whereas the impurities related to primary smelting are much lower, this leads to different flowsheet design and operating conditions. In fact, the pyrometallurgical process of secondary copper requires a higher level of impurity removal, together with special operations for gas cleaning in order to snatch toxic emissions such as dioxins, halogens, NOx, etc.

The first stage of e-waste treatment using the secondary copper route is smelting the feed materials under reducing conditions. This usually happens in a reduction furnace, and the product is called black copper. Black copper is a middle product that then can be further purified by oxidizing to get a clean copper product. A mixture of copper scrap, electronic waste, slag, and coke is injected into the reduction furnace. Then, air is blown through the top nozzles, which causes the coke to burn and hence the smelting of other feed materials under reducing conditions.

The black copper that is generated during the reduction process has a large amount of impurities. Throughout the reduction stage, most of the impurities are separated from the liquid copper, while some of them are cut apart in the vapor phase (dust of metals and their oxides).

In the course of the oxidation of black copper, which could be conducted in the same furnace or in a separate oxidation furnace, some impurities such as Zn, Sn, and Pb are taken out in the form of their oxides into the slag phase. Some other impurities such as lead and tin are separated into the gas phase and go through the filtering plant [25,32].

In the fire-refining stage, air and a reducing agent such as hydrocarbon are added to remove any existing oxides. Air is blown through molten metals, which oxidizes the existing impurities and removes sulfur. In the fourth stage, which is the electrorefining stage, an electrolytic cell is employed to separate copper and other metals. Copper in the impure anode is dissolved into an electrolyte to be coated onto a copper cathode. Insoluble impurities such as precious metals (Ag and Au), platinum group metals, Sn, and some minor amount of Pb in the electrolyte dropped to the bottom of the cell as anode slimes. The rest of the impure elements such as As, Bi, Ni, Sb, and Fe are partly or completely solvable in the electrolyte. The main reason for that is that the electrochemical properties of these elements are much lower than those of the electrorefining cell voltage; hence, they are not able to plate onto the cell.

The next stage of the process is electrowinning. In the electrowinning process, impure copper is leached in a copper sulfate solution and pure copper is recovered as a result. Lead or titanium acts as an insoluble anode, and steel or copper sheets act as cathodes in the process. The removal of copper from the solution and its deposition on the cathode happens in a similar way as in the electrorefining process. In the precious metal refining stage, valuable metals (mainly Au and Ag) are separated into the slime or sludge and deposited on the electrolytic cell. Then, the sludge goes through additional processing to recover Ag and Au. The flow sheet of the selected e-waste processing route is shown in Figure 2.

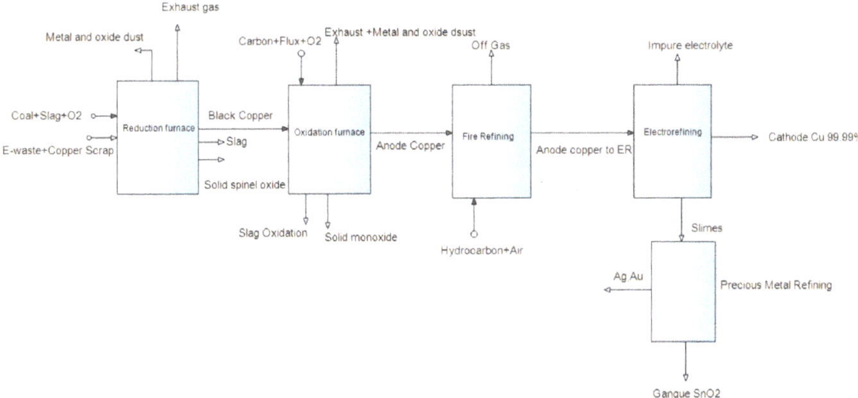

Figure 2. Flowsheet of the chosen process.

4. Methodology

The exergy balance of the proposed e-waste treatment process via secondary copper smelting has been developed using the HSC Chemistry thermochemical package (HSC and HSC Sim 7.1&8). This includes the modeling of black copper smelting process while adding 50% and 30% electronic waste of the total feed throughput into the feed.

The Fact-Sage 6.4 thermochemical package [26] is employed to calculate different element distribution in various flow streams (off gas, slag, and molten metal). The thermodynamic databases that exist in Fact-Sage include two main types, namely the compound database and the solution database. The compound database is employed for pure substance, and the solution database covered the adjusted parameters for solutions such as molten metals, liquid, or solid oxide solutions.

The Equilib module of Fact-Sage, which uses the Gibbs free-energy minimization algorithm, is used in this study. The Gibbs energy minimization mode calculates all of the equilibrium phase transitions by employing the user-defined compound and solution data. This module is basically utilized to predict the thermodynamically stable phases under assumed settings.

Databases

FactSage contains various optimized databanks for alloys solutions, slags, and liquid and solid oxides. In this study, optimized databases for copper-rich multicomponent systems and their correlated slags are adopted. "FactPS" and "FToxid" are used for the pure substances and liquid/solid oxide solutions, respectively. The FScopp database is used for copper alloy, and the SGnobl database used for evaluating the thermodynamic parameters of noble metals and their alloys.

The multicomponent structures consist of Cao, SiO_2, Fe_2O_3, FeO, and Al_2O_3 modified and used for the entire collection of compounds between 300–1700 °C. The liquid state oxides such as slag are analyzed by means of an amended quasichemical model that resolves the short-range ordering of compounds [25]. Cu_2O is assumed to an ideal solution and act as an elemental component of the slag. The behavior of liquid phase such as copper solution is signified by simplified polynomial expressions that are usable for a copper-rich liquid domain in temperature range of 300 to 1700 °C [25]. The equilibrium calculations are carried out for temperatures extending from 1100 to 1600 °C with partial pressures of oxygen changing from 10^{-7} to 10^{-10} atm [25]. For modeling purposes, difference oxygen partial pressures are set manually, and carbon is included respectively to reach the essential reducing condition objective.

For the oxidation stage of the process, the thermodynamic analysis is performed using multi-phase equilibrium calculations. This is done by including oxygen cumulatively (250 kg of oxygen added at each stage). This enables the tracing of slag formation, enhancing the purity of liquid copper, and carrying copper to the slag during the process's progression. The analysis of the oxidation stage happened at temperatures between 1000–1600 °C and pressure of 1 atm. Then, copper rich in a liquid state is separated and carried to the next stage of the process, which is fire refining. The equilibrium in the fire-refining phase of the process is calculated between the Cu-rich loquacious phase, the gashouse phase (air), and the reducing agents (hydrocarbon), which ended up becoming anode grade copper. The copper amount produced at the end of this stage of the process contained some degree of impurities such as Sn, Ni, Ag, and Au, which are separated in the electrorefining process. The analysis of this stage is conducted using the data from the literature.

Regarding the precious metal refining stage, which is the final stage of the process, the databases have not yet been optimized. For this reason, modeling the comportment of precious metals and aspects such as the partition ratio of precious metals in the slag and in the final product are carried out by gathering information from industry and available information in the literature.

The outcomes of the present analysis reported in this study merely rely on the thermodynamic calculations, and hence do not include the kinetics parameters that can have a significant influence on the total metal recovery in the actual process.

Through the thermodynamic analysis, it is estimated that temperatures of more than 1300 °C and an oxygen fractional pressure of 10^{-8} are needed for the proposed e-waste processing route [25]. Carbon that exists in the e-waste provides supplementary heat and reductant (such as CO) throughout the reduction phase of the process, and therefore can replace a proportion of the coke that is used as an input material. In this study, exergy analysis is conducted for two scenarios. In the "50% e-waste in the feed" scenario, 12 t/h of input material consisting of 5.94 t of e-waste (waste PCBs), 6 t of copper scrap, 0.08 tof coal, and 0.42 t of slag is fed into the reduction furnace.

The main feed materials composition that is employed in this study is presented in Tables 1 and 2. The compositions of these feed materials were given before in the work of Ghodrat et al. [25].

Table 1. Metal oxides, metallurgical coke, and slag composition utilized in the input, as publihsed by Ghodrat et al. [25].

Metal Oxide	Cu	Cu$_2$O	SnO$_2$	PbO	ZnO	NiO
wt%	70	7	5	8	5	5
Metallurgical coke elements	C	H$_2$O	S	Al$_2$O$_3$	FeO	
wt%	90	5	0.8	2	2.2	
slag	FeO	CaO	SiO$_2$			
wt%	45	17	38			

Table 2. Chosen compositions of PCB employed in the input, as published by Ghodrat et al. [25].

Element	Cu	Ag	Au	Al	Zn	Pb	Fe	Sn	Ni	Br	N	C	Al$_2$O$_3$	SiO$_2$
wt%	20.6	0.2	0.1	5	4	6	8.6	4	2	4	5	10	6	23.5

In "30% e-waste in the feed" scenario, 12 t/h input material consisting of 3.7 t of electronic waste (waste PCBs), 8.25 t of copper scrap, 0.08 t of coal, and 0.42 t of slag is fed into the reduction furnace. An overview of the entire modeling approach has been shown in Figure 3. In this gate-to-gate analysis, an exergy investigation has been conducted to assess the broadly used secondary copper recycling process by replacing part of the copper scrap with waste PCBs. The variable was the different percentage of e-waste in the feed material. A proposed flow sheet for the exergy balance of valuable metals recycling from waste printed circuit boards through secondary copper smelting is developed for two scenarios (50 wt% and 30 wt% e-waste in the feed) using HSC Chemistry and shown in Figures 4 and 5, respectively.

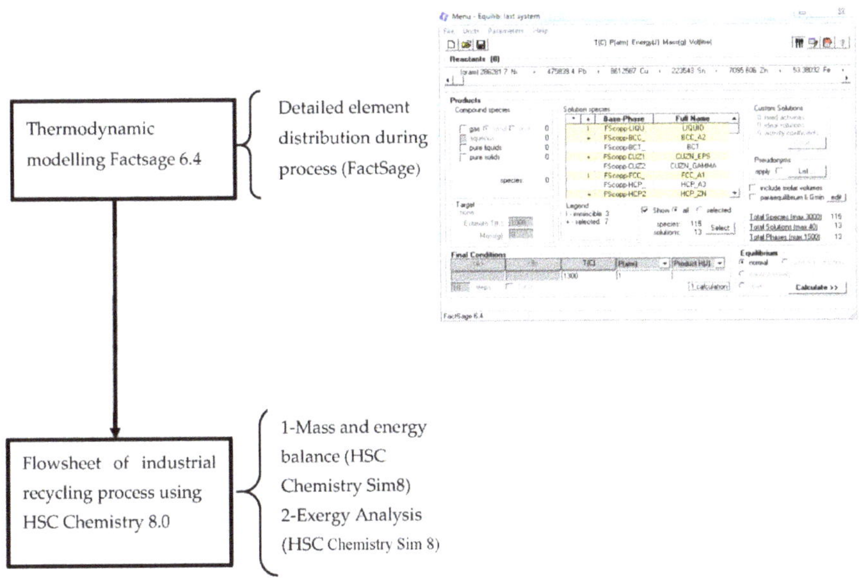

Figure 3. Flow chart of the procedure performing the exergy analysis.

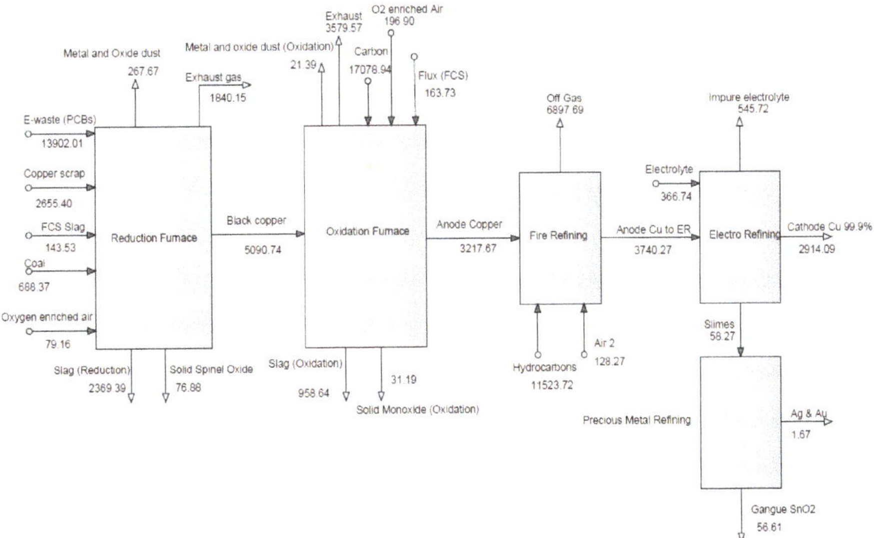

Figure 4. Total exergy balance (kWh), 50% e-waste in the throughput.

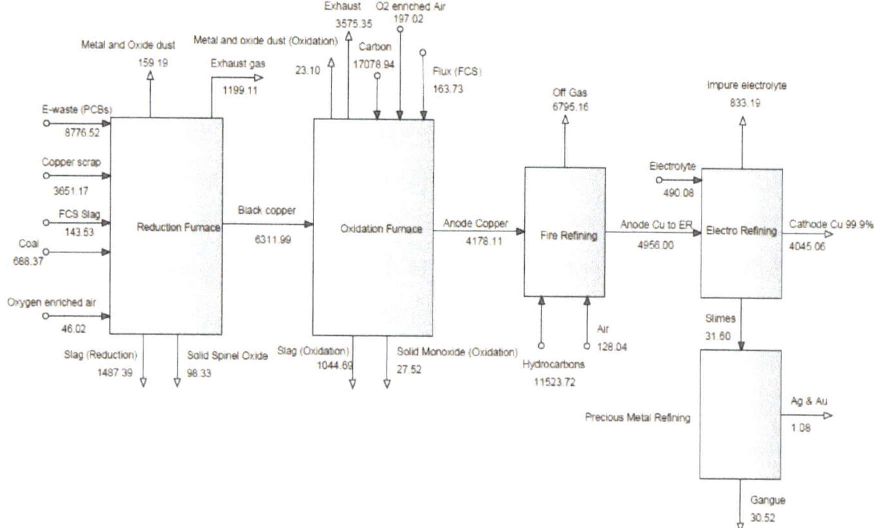

Figure 5. Total exergy balance (kWh), 30% e-waste in the throughput.

5. Results and Discussion

The element distribution in the output and input flows was estimated using a mixture of equilibrium modeling and data gathered from the relevant industries and the literature. As mentioned earlier, in this study, the Equilib module of Factsage using the Gibbs free-energy minimization mode and thermodynamic function of HSC chemistry is used for exergy calculation.

Specified compounds of chemical elements react to attain an equilibrium state. First, the reactants are defined; then, the compound and solution products are specified, and at the end, the final operational conditions of the process are set.

Based on the HSC Chemistry thermochemical database, the identified products in different phases include pure solids substance in the gas phase, spinel in the slag phase, and matte in the copper alloy solution phase. The distribution of elements in different phases was calculated by the "Dist" function of HSC Chemistry Sim 8 software and presented in Table 3.

Aluminum is a part of e-waste that is casted off to slag; however, this can lead to a higher slag viscosity and loss of the copper. About 50% of cobalt and a relatively high percentage of Ni are disseminated into copper phases.

In the fourth stage, which is electrorefining, Ni dissolves and is discarded in the electrolyte refinement route, although some proportion of the nickel concludes in the anode slime. Pb, Zn, and Sn are dropped in slag or off-gas, particularly in smelting and converting; a low percentage of these turn out to be in the blister copper. Pb may also be separated in the anode furnace using substances such as silica flux [30]. Pb, Zn, and Sn may also cut off in anode slime through electrorefining [30].

For the duration of smelting, Bi and As are scattered in the metal dusts phase, and can be recuperated via hydrometallurgical routes. The concentration of Sb and Bi needed to be controlled carefully in the electrolyte. The main reason is the presence of a high proportion of As to Sb and Bi; the molar ratio blocks the formation of floating slimes in the electrolyte.

Table 3. Distribution of elements between different phases.

Reduction			
Solid/Gas	**Gaseous**	**Solid**	**Liquid**
Metal and Oxide dust	**Exhaust Gas**	**Slag**	**Black Copper**
Zn, Pb, CuBr, PbBr(g) PbO, SnO, HBr(g), PbS Au, NiBr(g), NiBr$_2$, AgBr FeBr$_2$, InBr, GeO, SnS, Br ZnBr$_2$, PbBr$_2$, SnBr$_2$	CO$_2$(g), N$_2$(g) CO(g), H$_2$O(g) SO$_2$(g), H$_2$(g) SO(g)	Al$_2$O$_3$(l), SiO$_2$, CaO, FeO Fe$_2$O$_3$, PbO, ZnO, NiO, SnO(l), Cu$_2$O(l), FeS, PbS ZnS, NiS, Cu$_2$S, SnS, GeO$_2$ AgO, Ta	Ni, Pb, Cu(l) Sn, Zn, Fe Au, Ag, In Ge, Pd
		Solid Spinel Oxide	
		NiO, ZnO, Fe$_3$O$_4$ Al$_2$O$_3$	

Oxidation			
Solid/Gas	**Gaseous**	**Solid**	**Liquid**
Metal and Oxide Dust	**Exhaust Gas**	**Slag**	**Anode Copper to Fire Refining**
Zn, Pb, PbO, SnO, PbS, Au, GeO, SnS	CO$_2$(g), N$_2$(g) CO(g), H$_2$O(g) SO$_2$(g), H$_2$(g) SO(g), O$_2$(g)	FeO, CaO, SiO$_2$, Fe$_2$O$_3$ PbO, ZnO, NiO(l), SnO(l) Cu$_2$O(l), GeO$_2$, AgO	Ni, Pb, Cu(l), Sn, Zn Fe, Au, Ag, In, Ge, Pd
		Solid Monoxide	
		Fe$_3$O$_4$, CaO ZnO, NiO SnO$_2$	

Fire Refining			
Solid/Gas	**Gaseous**	**Solid**	**Liquid**
	Off Gas		**Anode Cu to Electro Refining**
	H$_2$(g), Co(g), H$_2$O(g) CO$_2$(g), Pb, Cu, Ag, Zn CH$_4$(g), SnO(g), GeO(g), Ni, Fe, N$_2$(g)		Ag, Au, C Cu(l), Fe, Ge In, Ni, Pb, Pd Sn, Zn

Electrorefining			
Solid/Gas	**Gaseous**	**Solid**	**Liquid**
		Impure Electrolyte	**Cathode Cu 99.9%**
		H$_2$O, H$_2$SO$_4$, Ni, As Ag, Au, C, Cu, Fe, Ge In, Pb, Pd, Sn, Zn	Ag, Au, C, Cu(l), Fe, Ge, In, Ni, Pb, Pd, Sn, Zn, S
			Slimes
			Ag, Au, C Cu, Fe, Ge, In Ni, Pb, Pd, Sn Zn, As

Precious Metal Refining			
Solid/Gas	**Gaseous**	**Solid**	**Liquid**
		Gangue	**Precious Metals**
		Ag, Au, C, Cu, Fe, Ge, In, Ni, Pb, Pd, Sn, Zn, As	Ag, Au

The results of the thermodynamic equilibrium calculations can help us understand on a chemical level the influence of different parameters such as the composition of materials, reducing agents, the oxidation process, pressure, or temperature on the composition of the resultant materials and the properties such as exergy quantity and losses during the studied process. Equilibrium calculations are employed to compute the steady phase accumulations based on the different solution composition as well as simulate the stable phase assembly, and hence evaluate the effect of the chemical composition on our recycling system. This thermodynamic calculation can also, in combination with an HSC model, be used to follow the changes (including changes in the chemical and total exergy) during different stages of the process. However, as many of the kinetic data regarding exergy calculation are not (yet) available, it is essential to comprehend the importance of conducting such studies. Thus far, thermodynamic equilibrium calculations have been applied mainly to mass and energy balance calculations, and not exergy evaluation. However, the approach is equally valid for primary copper smelting systems based on the primary input materials, and is expected to further the development of new waste materials and blends.

The values of the incoming and outgoing exergies of five process units involved in the e-waste treatment through the secondary copper smelting process are compared for the two considered scenarios (50% and 30% e-waste in the feed material) in Table 4. The exergy values of materials going into the reduction, oxidation, fire-refining, and electrorefining processes are shown in Tables 5–8, respectively.

The exergy values of molten copper, dust, slag, copper, and precious metal in sludge, off gases, impure electrolyte, and gangue are shown in Table 8 on a chemical element basis coming out of the reduction, oxidation, fire refining, electrorefining, and precious metal refining processes.

Table 4. Values of exergies of process units involved in electronic waste (e-waste) treatment through the copper recycling process.

	Inlet Exergies			Outlet Exergies		
	Source	Exergy (Total) (kWh)-50%	Exergy (Total) (kWh)-30%	Source	Exergy (Total) (kWh)-50%	Total Exergy (kWh)-30%
Reduction	E-waste	13,902	8776	Metal and oxide dust	228.33	129
	Copper scrape FCS	2655	3651	Exhaust gas	791.40	529
	(ferrous-calcia-silica) Slag	143	143	Slag	1135.29	716
	Coal	688	688	Solid spinel oxide	35.96	51
	O$_2$	79	46			
Oxidation	Carbon	17,078	17,078	Slag	542.83	610
	O$_2$ enriched air	196	197	Solid monoxide	12.91	12
	Flux (FCS)	163	164	Exhaust	958.36	958
				Metal and oxide dust	14.92	16
Fire refining	Hydrocarbon	14,701.78	11,523	Off gas	3956.67	3881
	Air	128.27	128			
Electrorefining	Electrolyte	360.23	482	Impure electrolyte	545.72	833
				Cathode Cu 99.9%	2914.09	3825
Precious metal refining	Slimes	58.27	31.60	Ag, Au	1.67	1.2
				Gangue	56.61	30.52

Table 5. Exergy values of materials injected into the reduction furnace.

Chemical Component	Chemical Exergy (kWh)-50%	Chemical Exergy (kWh)-30%	Chemical Component	Chemical Exergy (kWh)-50%	Chemical Exergy (kWh)-30%
Cu (E-waste stream)	722.49	456.12	Cu$_2$O	97.14	133.57
Cu (copper scarp stream)	2432.62	3344.85	SnO$_2$	22.29	30.65
Ag	1.52	0.96	PbO	37.42	51.45
Au	0.21	0.13	ZnO	29.22	40.18
Al	2715.80	1714.52	NiO	36.71	50.47
Zn	347.97	219.68	FeO (FCS slag stream)	96.54	96.54
Pb	119.12	75.20	CaO	45.20	45.20
Fe	977.73	617.26	SiO$_2$	1.79	1.79
Sn	306.84	193.71	C	683.16	683.16
Ni	136.34	86.08	H2O	0.06	0.06
Br	41.80	26.39	S	3.38	3.38
C	5636.05	3558.11	Al$_2$O$_3$	0.87	0.87
SiO$_2$	16.00	10.10	FeO (Coal stream)	0.90	0.90
In	0.13	0.08	O$_2$	69.07	40.16
Ge	0.25	0.16	N$_2$	10.09	5.87
Ta	0.18	0.11			
Pd	0.06	0.04			

Table 6. Exergy values of materials going into the oxidation furnace.

Chemical Component	Chemical Exergy (kWh)-50%	Chemical Exergy (kWh)-30%	Chemical Component	Chemical Exergy (kWh)-50%	Chemical Exergy (kWh)-30%
$O_2(g)$ (O_2 enriched Air stream)	171.81	172.05	Cu(l) (black copper stream)	3155.42	4056.26
$N_2(g)$ (O_2 enriched Air stream)	25.09	24.98	Sn (black copper stream)	593.65	604.12
FeO	55.29	55.29	Zn (black copper stream)	23.13	16.22
CaO	104.93	104.93	Fe (black copper stream)	4.39	5.36
SiO2	3.51	3.51	Au (black copper stream)	0.21	0.13
C	17,078.94	17,078.94	Ag (black copper stream)	1.49	0.96
Ni (black copper stream)	378.41	437.47	In (black copper stream)	0.04	0.04
Pb (black copper stream)	64.06	86.79	Ge (black copper stream)	0.15	0.12
			Pd (black copper stream)	0.06	0.04

Table 7. Exergy values of materials going into the fire-refining stage.

Chemical Component	Chemical Exergy (kWh)-50%	Chemical Exergy (kWh)-30%	Chemical Component	Chemical Exergy (kWh)-50%	Chemical Exergy (kWh)-30%
$CH_4(g)$	11,523.72	11,523.72	Zn	0.23	0.05
$O_2(g)$	76.25	76.11	Fe	0.03	0.02
$N_2(g)$	52.02	51.93	Au	0.21	0.13
Ni (Anode copper stream)	75.68	170.61	Ag	1.46	0.95
Pb	6.41	12.76	In	0.04	0.04
Cu(l)	2887.21	3731.76	Ge	0.09	0.05
Sn	78.36	44.71	Pd	0.06	0.04

Table 8. Exergy values of materials going into the electrorefining stage.

Chemical Component	Chemical Exergy (kWh)-50%	Chemical Exergy (kWh)-30%	Chemical Component	Chemical Exergy (kWh)-50%	Chemical Exergy (kWh)-30%
$H_2SO_4(l)$	191.04	242.81	C	113.66	113.66
$CuSO_4$	38.83	57.82	Cu	2887.21	3731.76
H_2O	30.03	37.64	Ge	0.09	0.05
Ni	48.58	77.25	In	0.04	0.04
As	51.75	66.36	Ni	22.70	170.61
Ag	1.46	0.95	Pb	0.64	7.58
Au	0.21	0.13	Pd	0.06	0.04
Zn	0.02	0.01	Sn	39.18	44.71

5.1. Effect of Metal and Oxide Dust

For the two proposed scenarios (30 wt% and 50 wt% e-waste in the feed material), the chemical exergy of waste emissions (metal and oxide dust and exhaust gas) and solid waste such as those in slag and solid monoxide are compared in Table 9. As can be seen from the data presented in Table 9, during the smelting of copper scrap and electronic waste in the reduction furnace, the chemical exergy of 822 MJ for metal and oxide dust is calculated for 5.03 t of produced copper, and 10 kg of gold and silver (for 50% e-waste in the feed scenario). The quality of the copper scrap melted in the reduction furnace mainly controls the produced dust composition. Common compositional ranges for metal and oxide dust obtained from thermodynamic modeling are presented in Table 9. Table 9 indicates that all of the process units considered in the proposed flowsheet follow a similar trend when the percentage of electronic waste in the feed is decreased.

Comparing the exergy going into the process units and exergy going out of the process unit's values presented in Tables 5–9, the total loss in exergy during the process is revealed to be 45.5%. The exergy losses in the reduction, oxidation, and fire-refining furnaces are caused by chemical reactions and heat transfer. It is of great interest to note that controlling the outlet temperature of liquid copper, slims, and impure electrolyte is of significant importance. The main reason is that the cathode copper contains an exergy value of 10,490 MJ as shown in Table 9.

Table 9. Exergy values of materials going out of reduction (a), oxidation (b), fire refining (c), and electrorefining (d) furnace on a chemical component basis.

(a)		
Output Streams	**Chemical Ex (kWh)**	
Metal and Oxide Dust	**Chem Ex (kWh)-50%**	**Chem Ex (kWh)-30%**
Total	228.33	129.82
$Zn(g)$	162.99	26.03
$Pb(g)$	13.09	59.61
$PbBr(g)$	1.20	5.45
PbO	0.23	1.05
SnO	3.82	0.19
$HBr(g)$	0.09	0.07
PbS	0.70	4.75
$FeBr_2$	0.40	2.27
$InBr$	0.06	0.03
GeO	0.03	0.01
SnS	1.87	0.13
Br	40.45	16.01
$ZnBr_2$	0.13	0.02
$PbBr_2$	0.05	0.20
$SnBr_2$	3.22	0.14
Exhaust Gas	**Chem Ex kWh-50%**	**Chem Ex (kWh)-30%**
Total	791.40	529.56
$CO_2(g)$	269.28	180.73
$N_2(g)$	12.21	7.20
$CO(g)$	508.42	341.24
$H_2O(g)$	0.05	0.05
$SO_2(g)$	0.71	-0.32
$H_2(g)$	0.72	0.66
$SO(g)$	0.01	-0.01
Slag (Reduction)	**Chem Ex (kWh)-50%**	**Chem Ex (kWh)-30%**
Total	1135.29	716.26
$Al_2O_3(l)$	491.01	288.13
$SiO_2(l)$	17.79	11.89
$CaO(l)$	45.20	45.20
$FeO(l)$	402.06	284.07
$Fe_2O_3(l)$	0.22	0.13
$PbO(s)$	47.41	30.95
$ZnO(s)$	29.24	40.85
$NiO(l)$	1.32	0.87
$SnO(l)$	3.35	4.70
$Cu_2O(l)$	96.99	9.21
$FeS(l)$	0.27	0
$PbS(l)$	0.06	0.04
$ZnS(l)$	0.08	0.11
$Cu_2S(l)$	0.06	0.01
$AgO(l)$	0.04	0
$Ta(s)$	0.18	0.11

Table 9. *Cont.*

Black Copper	Chem Ex (kWh)-50%	Chem Ex (kWh)-30%
Total	4221.02	5207.51
Ni	378.41	437.47
Pb	64.06	86.79
Cu(l)	3155.42	4056.26
Sn	593.65	604.12
Zn	23.13	16.22
Fe	4.39	5.36
Au	0.21	0.13
Ag	1.49	0.96
In	0.04	0.04
Ge	0.15	0.12
Pd	0.06	0.04

Solid Spinel Oxide	Chem Ex (kWh)-50%	Chem Ex (kWh)-30%
Total	35.96	51.91
NiO	2.54	1.86
ZnO	13.35	14.01
Fe_3O_4	10.06	7.54
Al_2O_3	10.02	28.50

	(b)	

Slag (Oxidation)	Chem Ex (kWh)-50%	Chem Ex (kWh)-30%
Total	542.83	610.36
FeO(l)	26.94	22.98
CaO(l)	104.93	104.93
SiO_2(l)	3.51	3.51
Fe_2O_3(l)	1.20	1.39
PbO(s)	7.73	13.09
ZnO(s)	1.78	1.27
NiO(l)	34.14	26.41
SnO(l)	241.99	290.88
Cu_2O(l)	120.58	145.89
AgO(l)	0.03	0.00

Exhaust	Chem Ex (kWh)-50%	Chem Ex (kWh)-30%
Total	958.36	958.24
CO_2(g)	818.75	818.75
N_2(g)	25.09	24.98
CO(g)	114.51	114.51

Metal and Oxide Dust (Oxidation)	Chem Ex (kWh)	Chem Ex (kWh)-30%
Total	14.92	16.43
Zn	1.16	0.25
Pb	6.73	4.39
PbO	5.07	4.41
SnO	1.95	7.37
GeO	0.01	0.01

Solid Monoxide (oxidation)	Chem Ex (kWh)-50%	Chem Ex (kWh)-30%
Total	12.91	11.74
Fe_3O_4	0.88	0.85
ZnO	0.02	0.05
NiO	6.93	9.79
SnO_2	5.07	1.05

Table 9. *Cont.*

Anode Copper to FR (Fire Refining)	Chem Ex (kWh)-50%	Chem Ex (kWh)-30%
Total	3049.79	3961.13
Ni	75.68	170.61
Pb	6.41	12.76
Cu(l)	2887.21	3731.76
Sn	78.36	44.71
Zn	0.23	0.05
Fe	0.03	0.02
Au	0.21	0.13
Ag	1.46	0.95
In	0.04	0.04
Ge	0.09	0.05
Pd	0.06	0.04

(c)

Off Gas	Chem Ex (kWh)-50%	Chem Ex (kWh)-30%
Total	3956.67	3881.38
$H_2(g)$	1334.30	1334.30
$Co(g)$	0.00	0.00
$H_2O(g)$	15.52	15.52
$CO_2(g)$	215.75	215.75
Pb	5.77	5.18
Zn	0.21	0.05
$CH_4(g)$	2258.65	2258.65
$SnO(g)$	21.44	0.00
Ni	52.98	0.00
Fe	0.03	0.00
$N_2(g)$	52.02	51.93

Anode Cu to ER (Electro Refining)	Chem Ex (kWh)-50%	Chem Ex (kWh)-30%
Total	3065.28	4069.55
Ag	1.46	0.95
Au	0.21	0.13
C	113.66	113.66
Cu	2887.21	3731.76
Ge	0.09	0.05
In	0.04	0.04
Ni	22.70	170.61
Pb	0.64	7.58
Pd	0.06	0.04
Sn	39.18	44.71
Zn	0.02	0.01

(d)

Impure Electrolyte	Chem Ex (kWh)-50%	Chem Ex (kWh)-30%
Total	545.72	833.19
H_2O	29.62	37.02
H_2SO_4	263.40	350.21
Ni	69.86	242.91
As	38.81	49.77
C	113.66	113.66
Cu	29.46	38.19
Pb	0.01	0.02
Zn	0.02	0.01
$O_2(g)$	0.88	1.32

Table 9. *Cont.*

Slimes	Chem Ex (kWh)	
Total	58.27	31.60
Ag	1.46	0.95
Au	0.21	0.13
Cu	2.95	0.00
Ge	0.09	0.05
In	0.04	0.04
Ni	0.71	2.48
Pb	0.63	7.50
Pd	0.06	0.04
Sn	39.18	0.00
As	12.94	16.59
Cathode Cu 99.9%	**Chem Ex (kWh)-50%**	**Chem Ex (kWh)-30%**
Total	2914.09	3825.45
Cu	2913.37	3776.96
Ni	0.71	2.48

5.2. Effect of Off-Gas Emission

Figure 6 shows the emission exergy based on the waste gas types for the e-waste processing through copper recycling with 50% and 30% e-waste (waste PCBs) in the feed during the period of one hour. As can be perceived from Figure 6, the values of emission exergy for exhaust gas (CO, NO_x, and SO_2) enhanced by 1.94 and 5.64 times in the oxidation and fire-refining stages in comparison to the reduction stage, respectively in a 50% e-waste scenario and by 3.13 and 5.66 times in a 30% e-waste scenario.

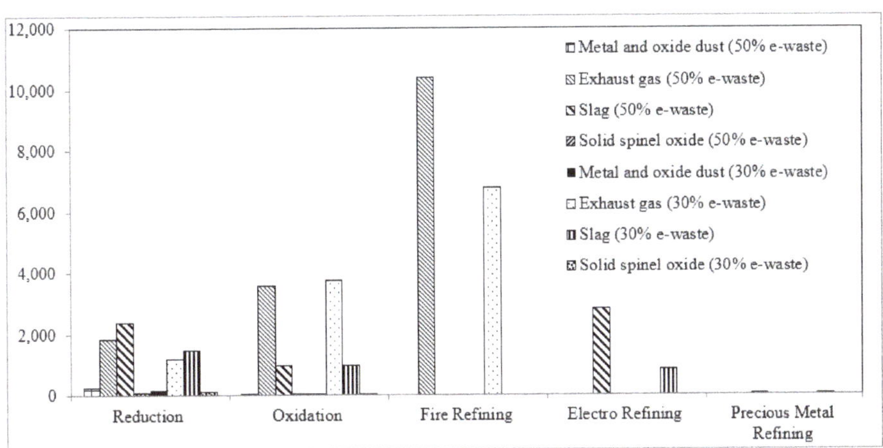

Figure 6. Emission exergy based on waste gas (50% and 30% e-waste in the feed).

The rate of the waste gas emission of each process unit is depicted in Figure 7. CO_2 has the foremost contribution with a rate enlarged from approximately 18% to more than 69% from the reduction to the oxidation stage in the 30% e-waste in the feed scenario, and from 21% to 63% in the 50% e-waste scenario, respectively.

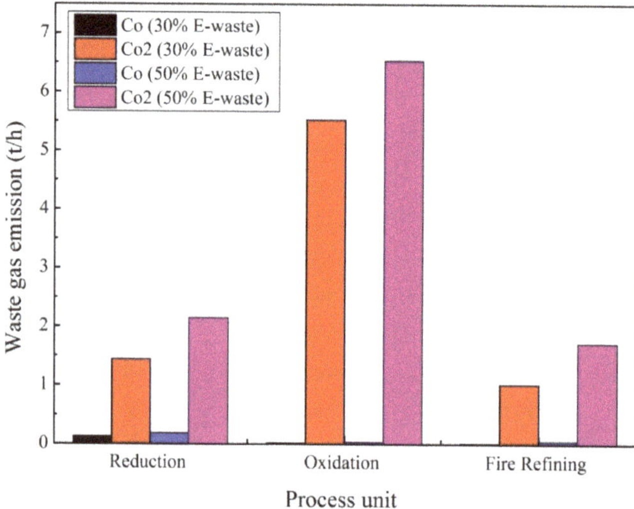

Figure 7. Waste gas emission rate (t/h).

Figure 8 made a comparison between the solid and gas waste emission generated by the two major process units (reduction and oxidation furnace) in e-waste treatment.

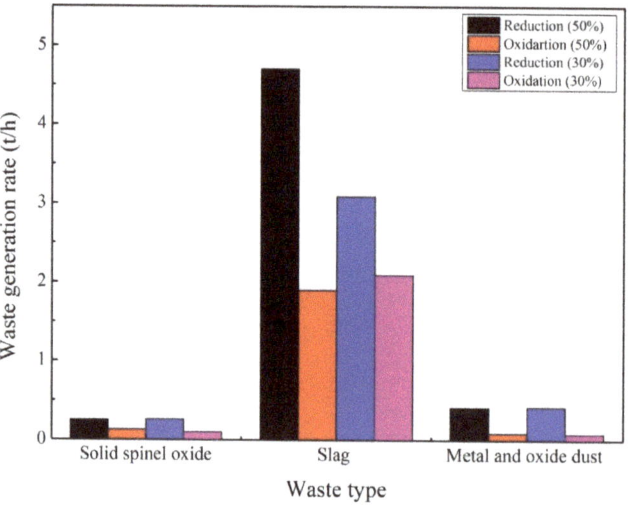

Figure 8. Solid waste generation rate (t/h).

As can be seen from Figure 8, it is estimated that almost 2200 t of solid spinel oxide dust, 41,000 t of slag, and 3600 t of metal and oxide dust was generated annually in the reduction furnace when the percentage of e-waste in the feed is 50 wt%.

From this, 3000 t was further recycled to recover Ni, Sn, etc., and slag was partially reused in a closed loop back to the reduction and oxidation furnaces, with the rest being landfilled.

The variation of the exergy loss for all of the process units involved in e-waste treatment through secondary copper smelting is presented in Figure 9. The oxidation stage is the key contributor for exergy loss followed by reduction and fire refining. As depicted in Figure 9, the exergy losses dropped down sharply from the reduction to oxidation furnaces, and then escalated and slowed down during the precious metal-refining stage.

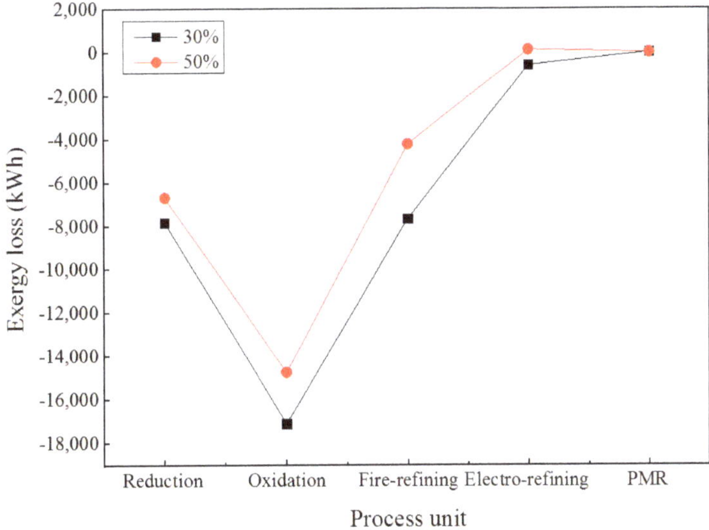

Figure 9. Exergy losses for different percentages of e-waste in the feed material.

The exergy efficiency of the five process units involved in the e-waste treatment through the secondary copper recycling process is shown in Figure 10. As can be seen from this figure, the precious metal refining (PMR) has the largest exergy efficiency followed by electrorefining and fire refining.

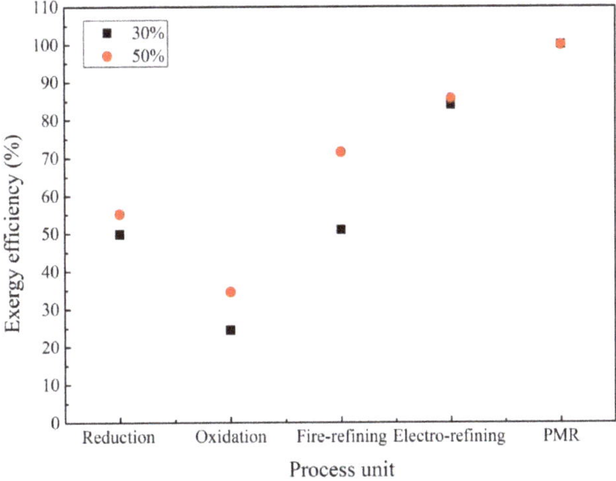

Figure 10. Exergy efficiency for different percentage of e-waste in the feed material.

5.3. Effect of Solid Waste in Slag

The exergy content of the two process streams was estimated and shown in Figure 11. As can be seen from this figure, the chemical exergy of the solid waste elements in slag has decreased noticeably by decreasing the e-waste percentage in the feed in the reduction process; however, the decrease is negligible in the oxidation process.

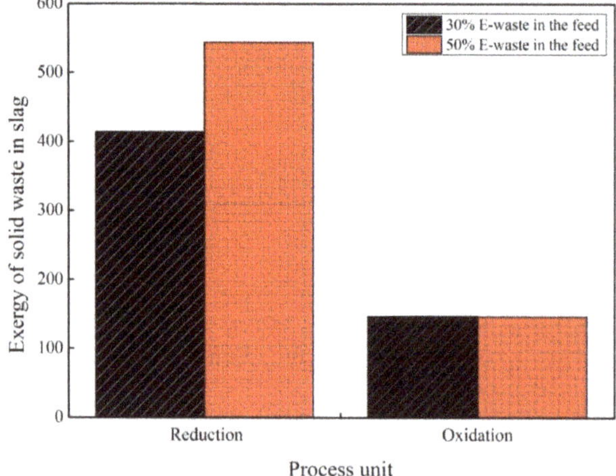

Figure 11. Effect of solid waste in slag on chemical exergy for two scenarios.

In order to understand the losses and performance of the proposed process for the two scenarios investigated in this study, a comparison has been made based on the environmental footprint and exergy losses of the flowsheets. The emphasis is more on the quantity of exhaust gas emitted from the process. From the results, it is evident that the "30% e-waste in-feed" scenario is superior to the "50% e-waste in-feed", as the amount of exhaust gas is about 34% less. On the other hand, as higher exergy loss is corresponded to a higher usage of coke and hence a higher environmental burden, once again, the 30% e-waste in the feed scenario is considered as a preferable option.

6. Conclusions

An exergy analysis for electronic waste treatment through a secondary copper-smelting route is thermodynamically modeled using HSC chemistry. The suggested technique has been employed to estimate the loss of exergy within the proposed precious metal recovery system, and helps calculate and assess the exergy efficiency for a different percentage of e-waste in the feed material. The results of the two case studies assessed here indicated that the value of emission exergy for exhaust gas (CO, NO_x, and SO_2) is enhanced in the oxidation and fire-refining stages in comparison to the reduction stage for both scenarios (50% and 30% e-waste in the feed). Moreover, by comparing the rate of the waste gas emission of each process unit, it has been revealed that CO_2 is the main contributor, with a sharp increasing rate from the reduction stage to the oxidation stage for both 30 wt% and 50 wt% e-waste in the material throughput. The variation of the exergy loss for all of the process units involved in e-waste treatment through secondary copper smelting has illustrated that the oxidation stage is the key contributor for exergy loss, followed by reduction and fire refining. In terms of exergy efficiency, precious metal refining (PMR) was found to have the highest exergy efficiency, followed by electrorefining and fire refining.

Further research has to be done to highlight the link between exergy analysis and economics of the metal recovery from waste PCB through the black copper smelting process. That makes the proposed recovery process more widely applicable for the recycling industry. Developing design tools to conduct a systematic study on the relations between exergy, material consumption, and life cycle assessment indicators is a second major area that needs further investigation. Understanding the human footprint and challenges of reaching a circular economy system is a topic of great interest. Further work has to be done to link the proposed recovery process to platinum group metal recovery using black copper smelting. This will allow an accurate assessment of the exergy destruction of compound product designs, including primary production, and hence accommodates the broad circular economy system.

Author Contributions: The research presented in this paper was a collaborative effort made by all the authors. All the authors contributed to the conceptualization, methodology, Software, validation, formal analysis, investigation, resources, data curation, writing and reviewing.

Funding: This research was funded by Centre for Infrastructure Engineering, Western Sydney University.

Conflicts of Interest: The authors declare no conflict of interest.

References

1. Rosen, M.A.; Dincer, I. On Exergy and Environmental Impact. *Int. J. Energy Res.* **1997**, *21*, 643–654. [CrossRef]
2. Szargut, J.; Morris, D.R.; Steward, F.R. *Exergy Analysis of Thermal, Chemical, and Metallurgical Processes*; Hemisphere Publishing Corporation: New York, NY, USA, 1988.
3. Edgerton, R.H. *Available Energy and Environmental Economics*; D.C. Heath: Toronto, Canada, 1982.
4. Ayres, R.U.; Ayres, L.W.; Martinas, K. *Eco-Thermodynamics: Exergy and Life Cycle Analysis*; INSEAD: Fontainebleau, France, 1996; Volume 19.
5. Ayres, R.U.; Ayres, L.W.; Martinás, K. Exergy, waste accounting, and life-cycle analysis. *Energy* **1998**, *23*, 355–363. [CrossRef]
6. Ayres, R.U.; Ayres, L.W.; Warr, B. Is the US economy dematerializing? Main indicators and drivers. In *Economics of Industrial Ecology: Materials*; Structural Change and Spatial Scales MIT Press: Cambridge, MA, USA, 2004.
7. Rosen, M.A.; Dincer, I.; Kanoglu, M. Role of exergy in increasing efficiency and sustainability and reducing environmental impact. *Energy Policy* **2008**, *36*, 128–137. [CrossRef]
8. Rosen, M.A.; Dincer, I. Exergy–Cost–Energy–Mass analysis of thermal systems and processes. *Energy Convers. Manag.* **2003**, *44*, 1633–1651. [CrossRef]
9. Daniel, J.J.; Rosen, M.A. Exergetic environmental assessment of life cycle emissions for various automobiles and fuels. *Exergy Int. J.* **2002**, *2*, 283–294. [CrossRef]
10. Dincer, I. Thermodynamics, Exergy and Environmental Impact. *Energy Sources* **2000**, *22*, 723–732. [CrossRef]
11. Dincer, I. The role of exergy in energy policy making. *Energy Policy* **2002**, *30*, 137–149. [CrossRef]
12. Dincer, I. Technical, environmental and exergetic aspects of hydrogen energy systems. *Int. J. Hydrogen Energy* **2002**, *27*, 265–285. [CrossRef]
13. Crane, P.; Scott, D.; Rosen, M. Comparison of exergy of emissions from two energy conversion technologies, considering the potential for environmental impact. *Int. J. Hydrogen Energy* **1992**, *17*, 345–350. [CrossRef]
14. Creyts, J.C. *Extended Exergy Analysis: A Tool for Assessment of the Environmental Impact of Industrial Processes*; University of California: Berkeley, CA, USA, 1998.
15. Gong, M.; Wall, G. On exergy and sustainable development—Part 2: Indicators and methods. *Exergy Int. J.* **2001**, *1*, 217–233. [CrossRef]
16. Ukidwe, N.U.; Bakshi, B.R. Thermodynamic Accounting of Ecosystem Contribution to Economic Sectors with Application to 1992 U.S. Economy. *Environ. Sci. Technol.* **2004**, *38*, 4810–4827. [CrossRef] [PubMed]
17. Ukidwe, N.U.; Bakshi, B.R. Industrial and ecological cumulative exergy consumption of the United States via the 1997 input–output benchmark model. *Energy* **2007**, *32*, 1560–1592. [CrossRef]
18. Ji, X.; Chen, G.Q.; Chen, B.; Jiang, M.M. Exergy-based assessment for waste gas emissions from Chinese transportation. *Energy Policy* **2009**, *37*, 2231–2240. [CrossRef]
19. Chen, G.Q. Scarcity of exergy and ecological evaluation based on embodied exergy. *Commun. Nonlinear Sci. Numer. Simul.* **2006**, *11*, 531–552. [CrossRef]
20. Dincer, I.; Rosen, M.A. Thermodynamic aspects of renewables and sustainable development. *Renew. Sustain. Energy Rev.* **2005**, *9*, 169–189. [CrossRef]
21. Ayres, R.U.; Ayres, L.W.; Martinas, K. Eco-Thermodynamics: Exergy and Life Cycle Analysis. 1996. Available online: https://flora.insead.edu/fichiersti_wp/inseadwp1996/96-04.pdf (accessed on 4 April 2019).
22. Amini, S.H.; Remmerswaal, J.A.M.; Castro, M.B.; Reuter, M.A. Quantifying the quality loss and resource efficiency of recycling by means of exergy analysis. *J. Clean. Prod.* **2007**, *15*, 907–913. [CrossRef]
23. Ignatenko, O.; van Schaik, A.; Reuter, M.A. Exergy as a tool for evaluation of the resource efficiency of recycling systems. *Miner. Eng.* **2007**, *20*, 862–874. [CrossRef]
24. Castro, M.B.G.; Remmerswaal, J.A.M.; Brezet, J.C.; Reuter, M.A. Exergy losses during recycling and the resource efficiency of product systems. *Resour. Conserv. Recycl.* **2007**, *52*, 219–233. [CrossRef]

25. Ghodrat, M.; Rhamdhani, M.A.; Khaliq, A.; Brooks, G.; Samali, B. Thermodynamic analysis of metals recycling out of waste printed circuit board through secondary copper smelting. *J. Mater. Cycles Waste Manag.* **2017**, *20*, 386–401. [CrossRef]
26. Bale, C.W.; Bélisle, E.; Chartrand, P.; Decterov, S.A.; Eriksson, G.; Hack, K.; Jung, I.-H.; Kang, Y.-B.; Melançon, J.; Pelton, A.D.; et al. FactSage thermochemical software and databases—Recent developments. *Calphad* **2009**, *33*, 295–311. [CrossRef]
27. Cornelissen, R.L. Thermodynamics and sustainable development. In *The Use of Exergy Analysis and the Reduction of Irreversibility*; Universiteit Twente: Enschede, The Netherlands, 1997; ISBN 9036510538.
28. Hait, J.; Jana, R.K.; Sanyal, S.K. Processing of copper electrorefining anode slime: A review. *Miner. Process. Extr. Metall.* **2009**, *118*, 240–252. [CrossRef]
29. Kulczycka, J.; Lelek, .; Lewandowska, A.; Wirth, H.; Bergesen, J.D. Environmental impacts of energy-efficient pyrometallurgical copper smelting technologies: The consequences of technological changes from 2010 to 2050. *J. Ind. Ecol.* **2016**, *20*, 304–316. [CrossRef]
30. Forsén, O.; Aromaa, J.; Lundström, M. Primary Copper Smelter and Refinery as a Recycling Plant—A System Integrated Approach to Estimate Secondary Raw Material Tolerance. *Recycling* **2017**, *2*, 19. [CrossRef]
31. Klemettinen, L.; Avarmaa, K.; O'Brien, H.; Taskinen, P.; Jokilaakso, A. Behavior of Tin and Antimony in Secondary Copper Smelting Process. *Minerals* **2019**, *9*, 39. [CrossRef]
32. Ghodrat, M.; Rhamdhani, M.A.; Brooks, G.; Masood, S.; Corder, G. Techno economic analysis of electronic waste processing through black copper smelting route. *J. Clean. Prod.* **2016**, *126*, 178–190. [CrossRef]

Article

A Risk Assessment Model of Coalbed Methane Development Based on the Matter-Element Extension Method

Wanqing Wang [1], Shuran Lyu [1,*], Yudong Zhang [2,*] and Shuqi Ma [1]

[1] School of Management Engineering, Capital University of Economics and Business, Beijing 100070, China; wwq199219@163.com (W.W.); shuqi_10@163.com (S.M.)

[2] Institute for Public Safety Research, Tsinghua University, Beijing 100084, China

* Correspondence: lsr22088@163.com (S.L.); z.yudong@foxmail.com (Y.Z.)

Received: 16 August 2019; Accepted: 14 October 2019; Published: 16 October 2019

Abstract: Coalbed methane development represents a complex system engineering operation that involves complex technology, many links, long cycles, and various risks. If risks are not controlled in a timely and effective manner, project operators may easily cause different levels of casualties, resource waste and property loss. To evaluate the risk status of coalbed methane development projects, this paper constructs a coalbed methane development risk assessment index system that consists of six first grade indexes and 45 second grade indexes. The weight of each index is calculated based on the structure entropy weight method. Then, a theoretical model for risk assessments of coalbed methane development is established based on the matter-element extension method. Finally, the model is applied to analyze a coalbed methane development project in the southern Qinshui Basin of China. The results show that the overall risk level of the coalbed methane development project is Grade II, indicating that the overall risk of the project is small, but the local risk of the project needs to be rectified in time. The assessment results are consistent with the actual operation of the project, indicating that the established risk assessment model has good applicability and effectiveness.

Keywords: coalbed methane development; risk assessment; structural entropy weight method; matter-element extension method

---

1. Introduction

With the rapid development of the global economy and the continuous advancement of urbanization, national dependence on black fossil energy, such as coal and oil, has increased. However, this increased dependence has caused excessive emissions to the atmosphere of toxic and harmful substances, such as carbon dioxide, sulfur dioxide, carbon monoxide and soot. These problems have caused environmental pollution problems, such as greenhouse effects, acid rain and haze, which have seriously affected the quality of life and health of residents. The above situation has received the attention of many actors, including the Chinese government. To achieve a balance between economic development and environmental protection, China's energy use structure is changing to a pattern of "high efficiency, low energy consumption, low pollution and low emissions" under the leadership of the government [1].

In this context, coalbed methane (CBM) has become a prominent resource. CBM is mainly composed of methane that is stored in coal seams, adsorbed on the surface of coal matrix particles, partially free in coal pores or dissolved in coal seam water hydrocarbon gas. It is an associated mineral resource of coal and unconventional natural gas, and it is also a clean and high-quality energy source. China is a large coal mining country that experiences frequent gas explosions [2]. Since 1949, 126 severe gas explosion accidents (defined as killing 30 or more people in one accident) have occurred, and 7502 people have died [3]. These accidents have not only caused serious casualties and property losses, but have also had a serious negative impact on society [4,5]. The development of CBM can not only improve environmental pollution [6], but also fundamentally reduce the methane content in the coal seam, which is beneficial to reducing the occurrence of coal mine gas explosions [7–9]. These advantages have driven CBM development as a component of the new energy industry and ushered it into a golden development period [1]. However, CBM development represents a complex system engineering operation that involves complex technology, many links, long cycles, and various risks. CBM development is highly susceptible to various risks, such as economic, legal, technological, and management risks, and risk control failure can result in casualties, resource waste and property losses. To avoid these risks and promote the steady development of the CBM industry, it is necessary to first understand and perceive the risks throughout the life cycle of CBM development via risk assessment research.

At present, achievements have been made in the research on CBM development risk assessment. Roadifer et al. [10] evaluated the future trends and risks of CBM development and identified the key factors affecting CBM reserves and productivity by combining experimental and mathematical-statistical methods. Senthi et al. [11] used Monte Carlo simulations and the hypercube model to evaluate the economic risks faced by the CBM industry. Chen et al. [12] also used the Monte Carlo simulation method to establish a risk transformation process model of the main uncertain factors in the CBM economy. Zhang et al. [13] determined the optimal index weights using the optimized combination entropy method and the triangular fuzzy number method and established a CBM development potential assessment model. Acquah-Andoh et al. [14] explored the best schemes for optimizing a company's revenue share for CBM development contracts based on factor analysis, discounted cash flow and parameter sensitivity analysis and found that the best scheme can distribute the economic risks of CBM development between governments and contractors. Luo et al. [15] used the net present value method to evaluate the economics of CBM production in China and found that the CBM price, productivity and operating costs are the three main factors affecting the economic feasibility of CBM development. Mares et al. [16] found that the uncertainty of adsorption capacity and desorption capacity were two important factors affecting the commercial development of CBM development, and their study provides a reference for the economic risk assessment of CBM development. Mu et al. [17] believed that three aspects are of great significance for avoiding CBM development risks: Pre-evaluation of CBM development, geological and gas reservoir engineering research and engineering technological innovation. Kirchgessne et al. [18] believed that safety and environmental factors also affect the economic benefits of CBM recovery. Su et al. [19] improved the discounted cash flow method by performing a hierarchical differentiation evaluation, staged evaluation and dynamic evaluation. It is also believed that production has the greatest impact on the economics of CBM development.

In summary, although scholars have made some achievements in the research on CBM development risks, the following shortcomings are observed:

(1)   Most studies focus on economic risk factor analyses of the national CBM industry. At present, risk assessment research for CBM development projects has not provided a reference for investors, insurance companies and CBM development operators.

(2)   Although several researchers have studied the risk of CBM development projects, the research depth was insufficient because the studies only determined the weight of the CBM risk assessment indicators. These studies used the weights to identify the main risk factors of CBM development, but failed to quantitatively measure the overall risk levels of CBM development projects and calculate the membership degree of each bottom index.

(3)   In terms of the construction of the indicator system, the above studies lack a systematic and comprehensive indicator selection process and did not cover the life cycle of CBM development, including geological exploration, drilling, gathering, and market applications. Therefore, the research results have a certain one-sidedness.

To fill the above gaps and achieve a reasonable assessment of the risk status of coalbed methane development projects, this paper will construct a risk assessment model for coalbed methane development. First, this paper seeks to construct a complete risk assessment index system for CBM development, including the laws, regulations and policies, resource characteristics, engineering technology and organizational management, according to the risk characteristics involved in the life cycle of CBM development.

Second, to scientifically determine the weight of each index, the weight calculation method should be chosen reasonably. The AHP, Delphi method and expert experience are commonly used methods to determine the index weights in risk assessment, and the AHP is the most widely used method [20]. However, the disadvantage of AHP is that if the index system contains a large number of indexes, it will greatly increase the workload of experts, which affects the acquisition of the judgment matrix, and thus, affects the accuracy of the weight calculation. Compared with the AHP, the structure entropy weight method (SEWM) can reduce a large amount of the computational workload and obtain more accurate results in the case of a large number of indexes. The SEWM combined the methods of subjective and objective assignments, as well as qualitative and quantitative analysis [21]. The main steps of the SEWM include: (1) Collecting experts' comments and forming the typical order; (2) analyzing the blind degree (uncertainty) of indexes; (3) normalized treatment of indexes; and (4) determining the index weight of each layer. Please refer to Section 2.2 for the specific calculation process.

Third, scientifically assess the risk of CBM development, and the assessment method should be rationally selected. The risk assessment method in this paper is based on the matter-element extension method (MEEM). Because MEEM is a method for multi-index comprehensive assessment and mainly based on the extrinsic matter-element model, extension set and correlation function theory [22], the method can judge the membership level of things according to different characteristics of the elements and less data and can avoid the randomness and subjectivity of the evaluation process to a certain extent. It is a combination of qualitative and quantitative methods. At present, this method has achieved good results in risk assessment in many fields, such as oil exploitation [23], tailings pond [24], and building fire [25]. The main steps of the MEEM include the following: (1) Determining the classical domains, joint domains, and matter elements; (2) calculating the correlation degrees; (3) assessing multi-level extension; and (4) classifying risk. Please refer to Section 2.3 for the specific calculation process. Finally, this paper will conduct a case study to verify the feasibility of the risk assessment model.

The main contents of this paper include the following parts: Section 1 introduces the research significance, research purposes, literature review and current deficiencies in the field of coalbed methane development risk research. Section 2 introduces the research steps of the article, constructs the risk assessment index system of coalbed methane development, introduces the calculation steps of the structural entropy weight method and uses this method to calculate the index weight, and then introduces the calculation process of the matter-element extension method. Section 3 conducts a case study to verify the validity of the evaluation model. Section 4 analyses and discusses the assessment results. Section 5 summarizes the conclusions of the article.

2. Methodology

This study follows four steps: (1) The key risk factors involved in the life cycle of CBM development were identified through literature analysis, field investigation, legal norm inquiry and expert consultations. Thus, the risk assessment index system of CBM development was determined; (2) The SEWM was used to determine the weight of each index; (3) Based on the MEEM, a theoretical model of risk assessment was constructed. The risk classification rules were determined; (4) A case study was implemented. The main research steps involved in this paper are shown in Figure 1.

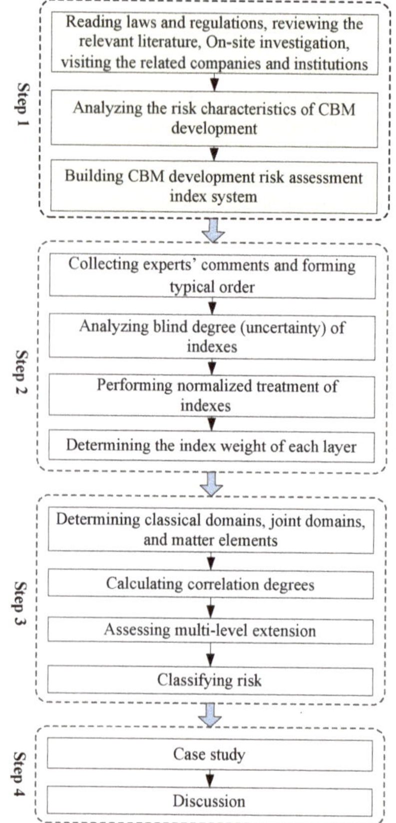

Figure 1. Steps of the present study. CBM, coalbed methane.

The four steps are described as follows:

Step 1: Build a targeted risk assessment index system for CBM development. First, the status quo of risk assessment and the key influencing risk factors of CBM development were achieved through on-site investigation, reading relevant literature and visiting the insurance company and a third-party risk assessment institution. On this basis, a preliminary risk assessment index system was established. Some experts from consulting organizations, management departments, and research institutions engaged in CBM development were invited to evaluate the preliminary index system. According to the experts' suggestions, this system mainly focused on the risks involved in geological resource exploration, drilling and drainage, gathering and transportation and market operations. Six main aspects, namely, laws and policies, resource characteristics, engineering technology, economic operation, organization and management, and safety and emergency protection, were covered. Finally, the first-level indexes were refined to the second-level indexes; thus, the risk assessment index system used in the CBM development was established.

Step 2: After the index system was determined, an index weight was assigned to all indexes in each layer. However, using the traditional analytic hierarchy process (AHP) method to assign the index weights would greatly increase the experts' workload. Therefore, this paper introduced a new method, the SEWM, to assign the index weights. This method combines the methods of subjective and objective assignments via qualitative and quantitative analyses. The detailed principles and application method of the SEWM will be explained in Section 2.2.1.

Step 3: For quantitative risk assessment, this paper established a theoretical model of risk assessment based on the MEEM. The model covers four parts: (1) Determining classical domains, sections, and matter elements; (2) calculating correlation degrees; (3) assessing multi-level extension; and (4) classifying risk. The detailed calculation steps will be described in Section 2.3.

Step 4: Case study. A CBM development project in the southern part of the Qinshui Basin in China was chosen for the case study to verify the feasibility of the assessment system.

2.1. Construction of the CBM Development Risk Assessment Index System

2.1.1. Principle

To reflect the risks of CBM development accurately and objectively, the scientific basis, guidance, operability, systematic process, comparability and comprehensiveness were considered to establish the index system in this paper.

2.1.2. Construction of Index System

Life cycle theory has been widely used in many fields, such as economics [26], environmental research [27], and management research [28]. The basic meaning of the life cycle can be understood as the whole process from "cradle to grave". For a product, the life cycle is the process of returning to nature from nature, which includes not only raw material collection and processing, but also the product storage, transportation and sales. According to the above definition, this paper divides the life cycle of CBM development into three main stages: Resource exploration, resource exploitation, and gathering and market operation. The risk characteristics of these three stages were analyzed, and the risk factors were summarized into the following six categories: (1) Risks of laws, regulations and policies; (2) risks of resource characteristics; (3) risks of engineering and technology; (4) risks of economic operations; (5) risks of organization and management; and (6) risks of safety and emergency protection.

Second, these six types of risks are used as first grade indexes, which are then refined to second grade indexes through an on-site investigation and a literature review [10–13,15–19,29–37] and related laws and regulations. The laws and regulations include the "Mineral Resources Law of the People's Republic of China" [38], "Coalbed Methane Industry Policy" [39], "Safety Production Law of the People's Republic of China" [40], and "Hazardous Chemicals Safety Management Regulations" [41]. Third, the index system was revised through expert consultation. Finally, the assessment index system of CBM development risk was determined (Figure 2), and it consisted of six first grade indexes and 45 s grade indexes. For a detailed index analysis, please refer to Appendix A.

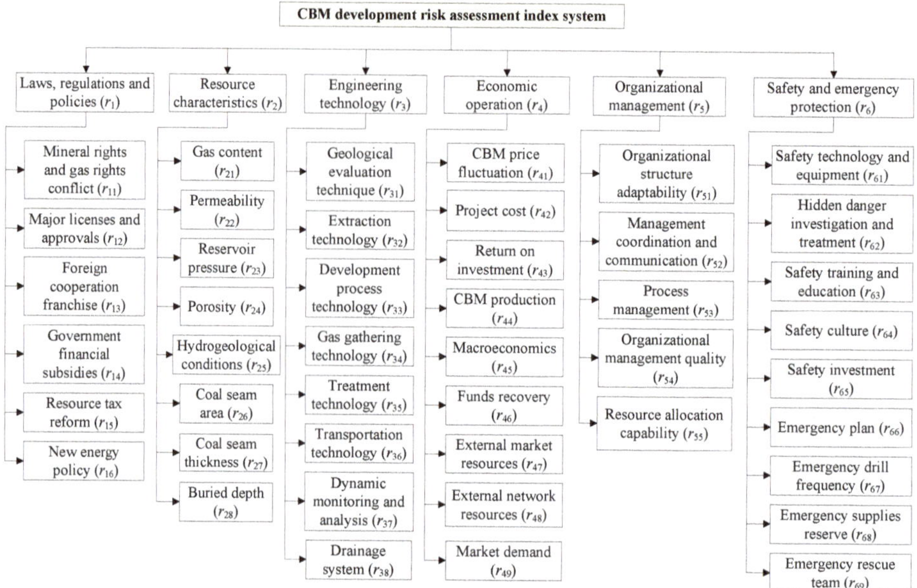

Figure 2. CBM development risk assessment index system.

2.2. Determination of Index Weights

To accurately calculate the weight of each index, a reasonable weight calculation method should be selected. The AHP, Delphi method and expert experience are commonly used methods to determine the index weights in risk assessment, and the AHP is the most widely used method [20]. Because many indexes are included in the proposed index system, the use of the AHP to determine the index weight would entail a large workload, which would not be conducive to acquiring the judgement matrix. Therefore, this paper introduced the SEWM to determine the index weight of the CBM development risk assessment system. The SEWM combines subjective and objective assignment methods, as well as qualitative and quantitative analyses. This method can reduce the calculation workload and achieve higher accuracy, especially for the many indexes in the CBM development risk assessment system.

2.2.1. Principle of the SEWM

The basic idea of the SEWM is to analyze the indexes of the assessment system and the interrelationship between them and then to classify the indexes into independent hierarchical grades. The execution steps are as follows.

(1)   Collection of experts' comments

Several experts were invited to complete the questionnaire form (Table 1) in accordance with the procedure and requirements of the Delphi method [42]. The experts ranked the importance of each index independently according to their own knowledge and experience. The indexes were ranked from high to low according to their importance; for example, mark "1" represented "most important", mark "2" represented "more important", mark "3" represented "important", and so on. Some indexes could be recognized as equally important, and the final rank of the indexes could be discussed by the experts.

Table 1. Collection of experts' comments.

	Index1	Index2	Index3	Index4	Index5	...
Expert1						
Expert2						
Expert3						
Expert4						
...						

(2) Blind degree (uncertainty) analysis

The potential deviation and uncertainty of the experts' comments on the index ranks might arise, due to noisy data. To eliminate noisy data and reduce uncertainty, the qualitative judgement conclusion from the experts should be statistically analyzed and addressed. To reduce the uncertainty of the experts' ranking, the entropy value was calculated by the entropy theory. The execution steps are shown below [21,43].

Supposing that k experts were invited to take the questionnaire survey, then k questionnaire forms would be returned, and every form would be recognized as an index set and marked as $R=\{r_1, r_2, \ldots r_k\}$; where r_i refers to the expert ranking array denoted by $\{a_{i1}, a_{i2}, \ldots a_{in}\}(i=1, 2, \ldots k)$ and $a_{i1}, a_{i2}, \ldots a_{in}$ can be any natural number from $\{1, 2, \ldots n\}$. As previously mentioned, "1" represents the highest level of importance. The index sort matrix obtained from the k table is shown as matrix A.

$$A = \begin{bmatrix} a_{11} & a_{12} & a_{13} & \cdots & a_{1n} \\ a_{21} & a_{22} & a_{23} & \cdots & a_{2n} \\ \cdots & \cdots & \cdots & \cdots & \cdots \\ a_{k1} & a_{k2} & a_{k3} & \cdots & a_{kn} \end{bmatrix} \tag{1}$$

where a_{ij} represents the ith expert's evaluation of the jth index.

The qualitative ranking result could be transformed into quantitative results by a membership function, which can be defined as follows:

$$\chi(I) = -\lambda p_n(I)\ln p_n(I), \tag{2}$$

where $p_n(I) = \frac{m-I}{m-1}, \lambda = \frac{1}{\ln(m-1)}$, which can be input into Equation (2):

$$\chi(I) = -\frac{1}{\ln(m-1)}\left(\frac{m-I}{m-1}\right)\ln\left(\frac{m-I}{m-1}\right). \tag{3}$$

Dividing both sides by $(m-I)/(m-1)$, assume that $1-\chi(I)/(m-I)/(m-1)=\mu(I)$. Then,

$$\mu(I) = \frac{\ln(m-I)}{\ln(m-1)}, \tag{4}$$

where I is defined as the qualitative ranking number of a certain index evaluated by an expert. For example, a set of qualitative ranking numbers 5, 2, 3, 4, 1 for the five indexes r_1, r_2, r_3, r_4, r_5 was evaluated by one expert. Thus, index r_5 was the most important, because $I = 1$. M is the transformation parameter and defined as $m = j + 2$, and j is the number of indexes.

The qualitative ranking number I is input into Equation (4) to obtain the quantitative transformation value of b_{ij}. $B_{ij} = \mu(a_{ij})$ is the membership degree of the qualitative ranking number I, and the matrix $B = (b_{ij})_{k*n}$ is defined as the membership degree matrix. A new parameter, average understanding degree b_j, was introduced to present the consistency degree of the evaluation of index r_j by k experts; its calculation is as follows:

$$b_j = \frac{b_{1j}+b_{2j}+\cdots+b_{kj}}{k}. \tag{5}$$

Blind understanding degree σ_j is defined as the uncertainty of the evaluation of index r_j by k experts,

$$\sigma_j = \left| \left\{ \left[\max\left(b_{1j}, b_{2j}, \cdots b_{kj}\right) - b_j \right] + \left[b_j - \min\left(b_{1j}, b_{2j}, \cdots b_{kj}\right) \right] \right\} / 2 \right|. \tag{6}$$

The global understanding degree X_j is defined as the degree of the evaluation of every index r_j by all k experts invited,

$$X_j = b_j\left(1 - \sigma_j\right). \tag{7}$$

(3) Normalized treatment

To obtain the weight of index r_j, Equation (7) needs further normalized treatment,

$$\omega_j = X_j / \sum_{j=1}^{k} X_j. \tag{8}$$

Obviously, $\omega_j > 0$ and $\sum_{j=1}^{k} \omega_j = 1$. The $\omega = (\omega_1, \omega_2, \ldots \omega_j)$ was expressed as the weight vector of the index set $R = (r_1, r_2, \ldots r_j)$.

2.2.2. Calculation of the Index Weight

In this study, a total of 12 experts who have worked in the CBM development industry for a long time, including three experts in CBM resource exploration, three experts in CBM mining technology, two experts in coal economy, two experts in energy policy, and two managers of CBM development enterprises. To more comprehensively formulate risk assessment criteria from multiple perspectives, the experts were randomly divided into four groups, each with three persons. In this way, the diverse understandings of various research fields could be fully explored, and different knowledge and experience could be used to perform a qualitative evaluation of each assessment index.

The weight determination of the indexes in the first grade was taken as an example:

(1) The rank results from the four groups of experts were collected in Table 2:

(2) The obtained rank matrix A:

$$A = \begin{bmatrix} 4 & 1 & 2 & 3 & 6 & 5 \\ 3 & 1 & 1 & 2 & 5 & 4 \\ 3 & 1 & 2 & 3 & 5 & 4 \\ 4 & 2 & 1 & 3 & 6 & 5 \end{bmatrix}$$

(3) The calculated membership matrix B was based on Equation (4) and rank matrix A, and m was set as 8.

$$B = \begin{bmatrix} 0.712 & 1.000 & 0.921 & 0.827 & 0.356 & 0.565 \\ 0.827 & 1.000 & 1.000 & 0.921 & 0.565 & 0.712 \\ 0.827 & 1.000 & 0.921 & 0.827 & 0.565 & 0.712 \\ 0.712 & 0.921 & 1.000 & 0.827 & 0.356 & 0.565 \end{bmatrix}$$

(4) The average degree of understanding of a particular dimension from all experts:

$$b_j = \frac{b_{1j} + b_{2j} + b_{3j} + b_{4j}}{4} = (0.770, 0.980, 0.960, 0.851, 0.460, 0.638).$$

(5) Based on the previous results and Equations (5) and (6), the blind understanding degree σ_j for the indexes from all experts could be obtained. Then, the evaluation vector X could be calculated according to the blind understanding degree σ_j and Equation (7). Finally, the weight of each index could be achieved by the normalized treatment method. The calculated result of each

parameter is shown in Table 3. Similarly, the weight distribution of the second grade indexes can be obtained by refining. The index distribution is shown in Table 4. The detailed calculation processes are shown in Appendix B (Tables A2–A7.).

Table 2. Expert ranking results.

	Index r_1	Index r_2	Index r_3	Index r_4	Index r_5	Index r_6
Group *A*	4	1	2	3	6	5
Group *B*	3	1	1	2	5	4
Group *C*	3	1	2	3	5	4
Group *D*	4	2	1	3	6	5

Table 3. Weight distribution of the first grade index.

	Index r_1	Index r_2	Index r_3	Index r_4	Index r_5	Index r_6
Group A	4	1	2	3	6	5
Group B	3	1	1	2	5	4
Group C	3	1	2	3	5	4
Group D	4	2	1	3	6	5
Average cognition degree b_j	0.770	0.980	0.960	0.851	0.460	0.638
$b_j max$	0.827	1.000	1.000	0.921	0.565	0.712
$b_j max - b_j$	0.057	0.020	0.040	0.070	0.104	0.074
$b_j min$	0.712	0.921	0.921	0.827	0.356	0.565
$b_j - b_j min$	0.057	0.059	0.040	0.023	0.104	0.074
σ_j	0.057	0.040	0.040	0.047	0.104	0.074
$1 - \sigma_j$	0.943	0.960	0.960	0.953	0.896	0.926
X_j	0.726	0.941	0.922	0.811	0.412	0.591
Weight	0.165	0.214	0.209	0.184	0.094	0.134

Table 4. Index weights and actual scores.

First Grade Indexes (r_i)	Weight (ω_i)	Second Grade Indexes (r_{ik})	Weight (ω_{ik})	Total Score	Actual Score (v_{ik})
Laws, regulations and policies (r_1)	0.165	Mineral rights and gas rights conflict (r_{11})	0.226	100	91
		Major licenses and approvals (r_{12})	0.235	100	88
		Foreign cooperation franchise (r_{13})	0.113	100	48
		Government financial subsidies (r_{14})	0.188	100	83
		Resource tax reform (r_{15})	0.118	100	86
		New energy policy (r_{16})	0.120	100	84
Resource characteristics (r_2)	0.214	Gas content (r_{21})	0.160	100	79
		Permeability (r_{22})	0.173	100	82
		Reservoir pressure (r_{23})	0.160	100	89
		Porosity (r_{24})	0.124	100	76
		Hydrogeological conditions (r_{25})	0.124	100	86
		Coal seam area (r_{26})	0.075	100	74
		Coal seam thickness (r_{27})	0.067	100	81
		Buried depth (r_{28})	0.118	100	87
Engineering technology (r_3)	0.209	Geological evaluation technique (r_{31})	0.170	100	86
		Extraction technology (r_{32})	0.158	100	74
		Development process technology (r_{33})	0.134	100	77
		Gas gathering technology (r_{34})	0.121	100	82
		Treatment technology (r_{35})	0.111	100	81
		Transportation technology (r_{36})	0.073	100	68
		Dynamic monitoring and analysis (r_{37})	0.098	100	78
		Drainage system (r_{38})	0.136	100	75

Table 4. *Cont.*

First Grade Indexes (r_i)	Weight (ω_i)	Second Grade Indexes (r_{ik})	Weight (ω_{ik})	Total Score	Actual Score (v_{ik})
		CBM price fluctuation (r_{41})	0.087	100	83
		Project cost (r_{42})	0.147	100	78
		Return on investment (r_{43})	0.134	100	76
Economic operation (r_4)	0.184	CBM production (r_{44})	0.145	100	72
		Macroeconomics (r_{45})	0.086	100	84
		Funds recovery (r_{46})	0.104	100	86
		External market resources (r_{47})	0.086	100	56
		External network resources (r_{48})	0.057	100	82
		Market demand (r_{49})	0.155	100	78
		Organizational structure adaptability (r_{51})	0.270	100	85
Organizational management (r_5)	0.094	Management coordination and communication (r_{52})	0.181	100	79
		Process management (r_{53})	0.205	100	73
		Organizational management quality (r_{54})	0.115	100	82
		Resource allocation capability (r_{55})	0.229	100	74
		Safety technology and equipment (r_{61})	0.151	100	75
		Hidden danger investigation and treatment (r_{62})	0.133	100	78
		Safety training and education (r_{63})	0.133	100	56
Safety and emergency protection (r_6)	0.134	Safety culture (r_{64})	0.062	100	72
		Safety investment (r_{65})	0.141	100	82
		Emergency plan (r_{66})	0.098	100	74
		Emergency drill frequency (r_{67})	0.088	100	54
		Emergency supplies reserve (r_{68})	0.095	100	63
		Emergency rescue team (r_{69})	0.100	100	76

2.3. Assessment Model Construction

The goal of the MEEM is to use the degree of association of the extension set to determine the assessment level of the matter element feature. The MEEM is a method for comprehensive multi-index assessments, and it is mainly based on the extrinsic matter-element model, extension set and correlation function theory [22]. This method can judge the subordinate level of items according to their different characteristics with low data requirements and can avoid the randomness and subjectivity of the evaluation process to a certain extent. The MEEM combines qualitative and quantitative analyses and has achieved good results in risk assessment in many fields, such as oil exploitation [23], tailings pond [24], and building fire [25]. Based on the above advantages, this method is applied to research CBM development risk assessments.

The MEEM includes the following steps [44,45]: (1) According to the development of things and relevant reference materials, the characteristics of things are analyzed, the things are divided into several grades according to certain rules, the numerical range of each level is clarified, and a multi-index MEEM is established; (2) Using the correlation function to calculate the degree of association between the things to be evaluated and each assessment level; (3) Things have the highest degree of relevance to one of the levels, indicating that they are most consistent with that level. Next, the calculation process of the CBM development risk assessment model will be described in detail.

2.3.1. Determination of the Classical Domain, Joint Domain and Matter-Element Evaluation

The matter element uses the ordered triplet $M = \{C, R, V\}$ as the basic element to describe things, where C is the name of the thing, R is the name of the feature, and V is the value taken by C for R [25].

(1) Determining the classical domain M_j

Let M_j be the classic domain of matter-element M:

$$M_j = \left(U_j, R, V_j\right) = \begin{pmatrix} U_j & r_1 & \left(a_{j1}, b_{j1}\right) \\ & r_2 & \left(a_{j2}, b_{j2}\right) \\ & \vdots & \vdots \\ & r_i & \left(a_{ji}, b_{ji}\right) \end{pmatrix}, \tag{9}$$

where U_j is the j risk level in the risk level domain U, V_j is the range of assessment index set R about the risk level U_j, and a_{ji} and b_{ji} are the lower and upper limits of the index r_i at the jth risk level, respectively.

(2) Determining the joint domain M_c

Let M_c be the joint domain of matter-element M:

$$M_c = (U, R, V_c) = \begin{pmatrix} U & r_1 & (a_{c1}, b_{c1}) \\ & r_2 & (a_{c2}, b_{c2}) \\ & \vdots & \vdots \\ & r_i & (a_{ci}, b_{ci}) \end{pmatrix}, \tag{10}$$

where V_c is the range of evaluation index set R about the risk level domain U, and a_{ci} and b_{ci} are the lower and upper limits of the index r_i at all risk levels, respectively.

(3) Determining the matter-element evaluation M_i

Let M_i be the matter-element evaluation of matter-element M:

$$M_i = (S_i, R_i, V_i) = \begin{pmatrix} S_j & r_{i1} & v_{i1} \\ & r_{i2} & v_{i2} \\ & \vdots & \vdots \\ & r_{ip} & v_{ip} \end{pmatrix}, \tag{11}$$

where S_i is the ith first grade index to be evaluated, $R_i=\{ r_{i1}, r_{i2}, \dots, r_{ip}\}$ is the second grade assessment index set for S_i, r_{ip} is the pth second grade assessment index of the ith first grade index, and V_i is the value of the second grade index R_i for S_i.

2.3.2. Calculating the Correlation Degree

By introducing the concept of distance in classical mathematics, the correlation function of the second grade index r_{ik} of the CBM development risk assessment on the risk level U_j is established. Therefore, the correlation degree $K_j(r_{ik})$ of the kth second grade index in the ith first grade index with respect to the risk level U_j is determined.

$$K_j(r_{ik}) = \begin{cases} \dfrac{\rho(v_{ik}, V_j)}{\rho(v_{ik}, V_c) - \rho(v_{ik}, V_j)} & \rho(v_{ik}, V_c) - \rho(v_{ik}, V_j) \neq 0 \\ -\rho(v_{ik}, V_j) - 1 & \rho(v_{ik}, V_c) - \rho(v_{ik}, V_j) = 0 \end{cases}, \tag{12}$$

where $\rho(v_{ik}, V_j) = \left| v_{ik} - \frac{a_{ji}+b_{ji}}{2} \right| - \frac{1}{2}(b_{ij} - a_{ji})$, $\rho(v_{ik}, V_c) = \left| v_{ik} - \frac{a_{ci}+b_{ci}}{2} \right| - \frac{1}{2}(b_{cj} - a_{ci})$, a_{ci} and b_{ci} are the lower and upper limits of the index r_i at all risk levels, respectively, and v_{ik} is the expert score for the second grade index r_{ik}.

2.3.3. Multi-Level Extension Assessment

(1) Primary assessment
Calculate the correlation matrix $K(r_i)$ of the first indexes for each risk level:

$$K(r_i) = (\omega_{i1}, \omega_{i2}, \cdots, \omega_{ip}) \begin{bmatrix} k_1(r_{i1}) & k_2(r_{i1}) & \cdots & k_m(r_{i1}) \\ k_1(r_{i2}) & k_2(r_{i2}) & \cdots & k_m(r_{i2}) \\ \vdots & \vdots & & \vdots \\ k_1(r_{ip}) & k_2(r_{ip}) & \cdots & k_m(r_{ip}) \end{bmatrix} = (k_j(r_i)), \tag{13}$$

where $\omega_i = (\omega_{ik})$ is the weight vector of the second grade indexes, and the calculation method is shown in Formulas (1)~(8); and $K(r_{ik}) = (k_j(r_{ik}))$ is the correlation degree matrix of the second grade indexes for each risk level.

(2) Secondary assessment

Determine the correlation degree matrix $K(S)$ of the CBM development safety for each risk level.

$$K(S) = (\omega_1, \omega_2, \cdots, \omega_j) \begin{bmatrix} k_1(r_1) & k_2(r_1) & \cdots & k_m(r_1) \\ k_1(r_2) & k_2(r_2) & \cdots & k_m(r_2) \\ \vdots & \vdots & & \vdots \\ k_1(r_p) & k_2(r_p) & \cdots & k_m(r_p) \end{bmatrix} = (k_j(S)), \tag{14}$$

where $\omega = (\omega_j)$ is the weight vector of the first grade indexes, and the calculation method is shown in formulas (1)~(8); a$^{\circledR}$$K(r) = (k(r_i))$ is the correlation degree matrix of the first grade indexes for each risk level.

(3) Determining the risk level

According to the principle of maximum membership degree, the risk level corresponding to the maximum correlation degree in the correlation degree matrix $K(S)$ of the CBM development for each risk level is the risk level of the assessment object. That is, when $maxk_j(S) = k_j(S)$, with $j = (1,2,3 \dots n)$, the risk level of the assessment object is at level j.

2.3.4. Risk Classification

To scientifically measure the risk of CBM development and ensure the systematic nature and accuracy of the assessment results, this paper divides the risk level of CBM development into five grades according to the actual CBM development situation and the risk classification rules in the literature [25,46], as shown in Table 5. Notably, when the established risk assessment model has been tested by a large number of empirical tests, the risk classification criteria can be corrected by data feedback to make the criteria more sensitive.

Table 5. Risk levels.

Risk Level	Score Range	Basic Characteristics
I	(85, 100]	System operation condition is very good, risk is very low
II	(70, 85]	System operation condition is good, risk is lower
III	(50, 70]	System operation condition is mediocre, risk is mediocre
IV	(25, 50]	System operation condition is poor, risk is higher
V	[0, 25]	System operation condition is very poor, risk is very high

3. Case Study

The southern part of the Qinshui Basin is one of the earliest areas for CBM exploration and development in China, and it has also attracted the most investment and research in CBM exploration and development in China [47]. The CBM storage conditions in this area are stable and have good development potential [48,49]. As a key breakthrough area for CBM exploration and development, many experts have carried out exploration and research work here, leading to the accumulation of a wealth of test and production data [50]. The experts have a deeper understanding of the characteristics of reservoir CBM accumulation, reservoir geological conditions and gas layer distribution. Therefore, this paper takes a CBM development project in this area as the research object and invited 5 experts, including 1 CBM exploration expert, 2 CBM mining technical experts, 1 energy policy expert and 1 project manager, to participate. Based on the engineering practice data of the project, the experts anonymously scored the actual operation status of each second grade index. The total score for each index is 100. In this paper, the actual scores of the second grade indexes of the CBM development project are obtained by calculating the average value, as shown in Table 4.

3.1. Determination of the Classical Domain, Joint Domain and Matter-Element Evaluation

Take the first grade index laws, regulations and policies (r_1) as an example to establish the matter-element M and determine its classical domain M_j, joint domain M_c and matter-element evaluation M_i. Similarly, classical domains, joint domains and matter-element evaluations of other first grade indexes can be obtained.

(1) Determining the classical domain M_j

The classical domains for each risk level are determined by Formula (9):

$$
M_\mathrm{I} = \begin{pmatrix} U_\mathrm{I} & r_{11} & (85,100) \\ & r_{12} & (85,100) \\ & r_{13} & (85,100) \\ & r_{14} & (85,100) \\ & r_{15} & (85,100) \\ & r_{16} & (85,100) \end{pmatrix}
\quad
M_\mathrm{II} = \begin{pmatrix} U_\mathrm{II} & r_{11} & (70,85) \\ & r_{12} & (70,85) \\ & r_{13} & (70,85) \\ & r_{14} & (70,85) \\ & r_{15} & (70,85) \\ & r_{16} & (70,85) \end{pmatrix},
$$

$$
M_\mathrm{III} = \begin{pmatrix} U_\mathrm{III} & r_{11} & (50,70) \\ & r_{12} & (50,70) \\ & r_{13} & (50,70) \\ & r_{14} & (50,70) \\ & r_{15} & (50,70) \\ & r_{16} & (50,70) \end{pmatrix}
\quad
M_\mathrm{IV} = \begin{pmatrix} U_\mathrm{IV} & r_{11} & (25,50) \\ & r_{12} & (25,50) \\ & r_{13} & (25,50) \\ & r_{14} & (25,50) \\ & r_{15} & (25,50) \\ & r_{16} & (25,50) \end{pmatrix},
$$

$$
M_\mathrm{V} = \begin{pmatrix} U_\mathrm{V} & r_{11} & (0,25) \\ & r_{12} & (0,25) \\ & r_{13} & (0,25) \\ & r_{14} & (0,25) \\ & r_{15} & (0,25) \\ & r_{16} & (0,25) \end{pmatrix},
$$

(2) Determining the joint domain M_c

The joint domain M_c is determined by Formula (10):

$$
M_c = \begin{pmatrix} U & r_{11} & (0,100) \\ & r_{12} & (0,100) \\ & r_{13} & (0,100) \\ & r_{14} & (0,100) \\ & r_{15} & (0,100) \\ & r_{16} & (0,100) \end{pmatrix}.
$$

(3) Determining the matter-element evaluation M_i

The matter-element evaluation M_1 of indexes r_{1i} is determined by Formula (11):

$$
M_1 = \begin{pmatrix} S_1 & r_{11} & v_{11} \\ & r_{12} & v_{12} \\ & \vdots & \vdots \\ & r_{1p} & v_{1p} \end{pmatrix} = \begin{pmatrix} R_1 & r_{11} & 91 \\ & r_{12} & 88 \\ & r_{13} & 48 \\ & r_{14} & 83 \\ & r_{15} & 86 \\ & r_{16} & 84 \end{pmatrix}.
$$

3.2. Calculating the Correlation Degree

Taking the correlation degree of the second grade index r_{11} of the first grade index r_1 as an example, the calculation steps of the correlation degrees of the second grade indexes are obtained by Formula (12).

$$\rho(v_{11},V_1)=|91-(85+100)/2|-(100-85)/2=-6;\ \rho(v_{11},V_2)=|91-(70+85)/2|-(85-70)/2=6;$$

$$\rho(v_{11},V_3)=|91-(50+70)/2|-(70-50)/2=21;\ \rho(v_{11},V_4)=|91-(25+50)/2|-(50-25)/2=41;$$

$$\rho(v_{11},V_5)=|91-(25+0)/2|-(25-0)/2=66;\ \rho(v_{11},V_c)=|91-(0+100)/2|-(100-0)/2=-9;$$

$$k_1(r_{11})=\rho(v_{11},V_1)/[\ \rho(v_{11},V_c)-\rho(v_{11},V_1)]=-6/(-9+6)=2;$$

$$k_2(r_{11})=\rho(v_{11},V_2)/[\ \rho(v_{11},V_c)-\rho(v_{11},V_2)]=6/(-9-6)=-0.4;$$

$$k_3(r_{11})=\rho(v_{11},V_3)/[\ \rho(v_{11},V_c)-\rho(v_{11},V_3)]=21/(-9-21)=-0.7;$$

$$k_4(r_{11})=\rho(v_{11},V_4)/[\ \rho(v_{11},V_c)-\rho(v_{11},V_4)]=41/(-9-41)=-0.82;$$

$$k_4(r_{11})=\rho(v_{11},V_5)/[\ \rho(v_{11},V_c)-\rho(v_{11},V_5)]=66/(-9-66)=-0.88;$$

Similarly, the correlation degrees of the second grade indexes under other first grade indexes can be calculated. The calculation results are shown in Table 6. The calculation process is shown in Appendix C.

Table 6. The relevance of second grade indexes.

First Grade Indexes	Second Grade Indexes	Relevance of Second Grade Indexes					Risk Level
		I	II	III	IV	V	
Laws, regulations and policies (r_1)	Mineral rights and gas rights conflict (r_{11})	2.000	−0.400	−0.700	−0.820	−0.880	I
	Major licenses and approvals (r_{12})	0.333	−0.200	−0.600	−0.760	−0.840	I
	Foreign cooperation franchise (r_{13})	−0.435	−0.314	−0.040	0.043	−0.324	IV
	Government financial subsidies (r_{14})	−0.105	0.133	−0.433	−0.660	−0.773	II
	Resource tax reform (r_{15})	0.077	−0.067	−0.533	−0.720	−0.813	I
	New energy policy (r_{16})	−0.059	0.067	−0.467	−0.680	−0.787	II
Resource characteristics (r_2)	Gas content (r_{21})	−0.222	0.400	−0.300	−0.580	−0.720	II
	Permeability (r_{22})	−0.143	0.200	−0.400	−0.640	−0.760	II
	Reservoir pressure (r_{23})	0.571	−0.267	−0.633	−0.780	−0.853	I
	Porosity (r_{24})	−0.273	0.333	−0.200	−0.520	−0.680	II
	Hydrogeological conditions (r_{25})	0.077	−0.067	−0.533	−0.720	−0.813	I
	Coal seam area (r_{26})	−0.297	0.182	−0.133	−0.480	−0.653	II
	Coal seam thickness (r_{27})	−0.174	0.267	−0.367	−0.620	−0.747	II
	Buried depth (r_{28})	0.182	−0.133	−0.567	−0.74	−0.826	I
Engineering technology (r_3)	Geological evaluation technique (r_{31})	0.770	−0.067	−0.533	−0.720	−0.813	I
	Extraction technology (r_{32})	−0.297	0.182	−0.133	−0.480	−0.653	II
	Development process technology (r_{33})	−0.258	0.438	−0.233	−0.540	−0.693	II
	Gas gathering technology (r_{34})	−0.143	0.200	−0.400	−0.640	−0.760	II
	Treatment technology (r_{35})	−0.174	0.267	−0.367	−0.620	−0.747	II
	Transportation technology (r_{36})	−0.347	−0.059	0.067	−0.360	−0.573	III
	Dynamic monitoring and analysis (r_{37})	−0.241	0.467	−0.267	−0.560	−0.707	II
	Drainage system (r_{38})	−0.286	0.250	−0.167	−0.500	−0.667	II
Economic operation (r_4)	CBM price fluctuation (r_{41})	−0.105	0.133	−0.433	−0.660	−0.773	II
	Project cost (r_{42})	−0.241	0.467	−0.267	−0.560	−0.707	II
	Return on investment (r_{43})	−0.273	0.333	−0.200	−0.520	−0.68	II
	CBM production (r_{44})	−0.317	0.077	−0.067	−0.440	−0.627	II
	Macroeconomics (r_{45})	−0.059	0.067	−0.467	−0.680	−0.787	II
	Funds recovery (r_{46})	0.077	−0.067	−0.533	−0.720	−0.813	I
	External market resources (r_{47})	−0.397	−0.214	0.158	−0.120	−0.413	III
	External network resources (r_{48})	−0.143	0.200	−0.400	−0.640	−0.760	II
	Market demand (r_{49})	−0.241	0.467	−0.267	−0.560	−0.707	II

Table 6. *Cont.*

First Grade Indexes	Second Grade Indexes	Relevance of Second Grade Indexes					Risk Level
		I	II	III	IV	V	
Organizational management (r_5)	Organizational structure adaptability (r_{51})	0.000	0.000	−0.500	−0.700	−0.800	I, II
	Management coordination and communication (r_{52})	−0.222	0.400	−0.300	−0.580	−0.720	II
	Process management (r_{53})	−0.308	0.125	−0.100	−0.460	−0.640	II
	Organizational management quality (r_{54})	−0.142	0.200	−0.400	−0.640	−0.760	II
	Resource allocation capability (r_{55})	−0.297	0.182	−0.133	−0.480	−0.653	II
Safety and emergency protection (r_6)	Safety technology and equipment (r_{61})	−0.286	0.250	−0.167	−0.500	−0.667	II
	Hidden danger investigation and treatment (r_{62})	−0.241	0.467	−0.267	−0.560	−0.707	II
	Safety training and education (r_{63})	−0.397	−0.241	0.158	−0.120	−0.413	III
	Safety culture (r_{64})	−0.317	0.077	−0.067	−0.440	−0.627	II
	Safety investment (r_{65})	−0.143	0.200	−0.400	−0.640	−0.760	II
	Emergency plan (r_{66})	−0.297	0.182	−0.133	−0.480	−0.653	II
	Emergency drill frequency (r_{67})	−0.403	−0.258	0.095	−0.080	−0.387	III
	Emergency supplies reserve (r_{68})	−0.373	−0.159	0.233	−0.260	−0.507	III
	Emergency rescue team (r_{69})	−0.273	0.333	−0.200	−0.520	−0.680	II

3.3. Multi-Level Extension Assessment

(1) Primary assessment

Taking the first grade index r_1 as an example, the correlation degree matrix $K(r_1)$ is determined according to the weight calculation result in Table 4 and Formula (13).

$$K(r_1) = (\omega_{11}, \omega_{12}, \omega_{13}, \omega_{14}, \omega_{15}, \omega_{16})(k_j(r_1)) =$$

$$(0.226, 0.235, 0.113, 0.188, 0.118, 0.120)\begin{bmatrix} 0.200 & -0.400 & -0.700 & -0.820 & -0.880 \\ 0.333 & -0.200 & 0.600 & -0.760 & -0.840 \\ -0.435 & -0.314 & -0.040 & -0.043 & -0.324 \\ -0.105 & 0.133 & -0.433 & -0.660 & -0.773 \\ 0.077 & -0.067 & -0.533 & -0.720 & -0.813 \\ -0.059 & 0.067 & -0.467 & -0.680 & -0.787 \end{bmatrix}$$

$$= (0.463, -0.148, -0.504, -0.650, -0.769)$$

In the same way, the correlation degree matrixes of other first grade indexes are calculated. The detailed calculation processes are shown in Appendix B.

$$K(r_2)=(-0.006, 0.105, -0.411, -0.647, -0.765); K(r_3)=(-0.075, 0.205, -0.276, -0.566,$$
$$-0.711); K(r_4)=(-0.204, 0.200, -0.260, -0.540, -0.694); K(r_5)=(-0.188, 0.163, -0.286,$$
$$-0.572, -0.714); K(r_6)=(-0.295, 0.114, -0.103, -0.414, -0.610).$$

(2) Secondary assessment

According to formula (14), the comprehensive correlation degree matrix of the research object for each risk level is determined.

$$K(S) = (\omega_1, \omega_2, \omega_3, \omega_4, \omega_5, \omega_6)K_i(r_i) =$$

$$(0.165, 0.214, 0.209, 0.184, 0.094, 0.134)\begin{bmatrix} 0.463 & -0.148 & -0.504 & -0.650 & -0.769 \\ -0.006 & 0.105 & -0.411 & -0.647 & -0.765 \\ -0.075 & 0.205 & -0.276 & -0.566 & -0.711 \\ -0.204 & 0.200 & -0.260 & -0.540 & -0.694 \\ -0.188 & 0.163 & -0.286 & -0.572 & -0.714 \\ -0.295 & 0.114 & -0.103 & -0.414 & -0.610 \end{bmatrix}$$

$$= (-0.035, 0.108, -0.317, -0.573, -0.716)$$

(3) Determining the risk level

According to the principle of maximum membership degree, the risk level corresponding to the maximum correlation degree in correlation degree matrix $K(S)$ of CBM development for each risk level is the risk level of the assessment object. Because $maxk_j(S)=0.108=K_2(S)$, the risk level of the CBM development project is Grade II.

4. Results and Discussion

(1). The correlation degree of comprehensive risk of the research object is 0.108, and the overall risk level of the CBM development project is judged to be Grade II according to the principle of maximum membership degree. This finding shows that the overall risk of the project is small and within an acceptable range, but the local risk of the project needs to be rectified in a timely manner.

(2). According to the calculation results of the correlation degree of the second grade indexes in Table 6, it is possible to determine the links with higher risks in the second grade indexes. The following is an analysis of the indexes with risk levels greater than Grade II.

(1) According to the investigation, the operators of the project did not obtain the right to cooperate with foreign countries, which indicates that the project cannot be assisted by foreign companies in terms of technology and management. Therefore, the project is at a higher risk in foreign cooperation franchises.

(2) The transportation technology of this project is relatively backwards, and mainly relies on tank trucks for CBM transportation, and there is no long-distance pipeline network for the system. Therefore, the transportation technology risk of this project is high. It is recommended that the project operator regularly check the reliability of the tank truck equipment and conduct safety training for the tank truck driver.

(3) Because the project operators are affiliated with private enterprises, and the enterprises have been established only for a short period, a well-known corporate image or reputation in the CBM industry have not been established. For these reasons, the project's external market resource risk is high.

(4) The project is not doing well in safety training and education. According to the survey, not all employees have participated in safety training, which will cause high risks to the daily operation of the project. CBM leakage and explosion accidents caused by human error are not uncommon; therefore, project operators should pay more attention to this aspect.

(5) The number of daily emergency drills of the project is not up to standard. During the investigation of the emergency drill record, it was found that the project conducted only one accident emergency drill every year, which did not meet the standards of China' China's Safe Production Law stipulates that emergency drills be performed at least once every six months for firms involved in the production, filling, storage, supply and sales of flammable and explosive chemicals and the emergency plan is to be continuously improved in light of the actual situation. Therefore, to reduce the project risk, the project operators should add at least one more accident emergency drill every year.

(6) The project's emergency supplies were not adequately prepared and did not include gas masks and explosion-proof emergency lights. The main component of CBM is methane, which is a flammable, explosive, toxic and corrosive gas. If a CBM leak accident occurs during the production process, rescue personnel are required to have a gas mask; otherwise, death, due to suffocation, is likely. Moreover, project operators should be equipped with a sufficient number of explosion-proof emergency lights. Other lighting equipment may cause methane explosion accidents, due to static electricity during use.

(7) The risks of other second grade indexes are very small, and their risk levels reached the first or second level. First, the risks of laws, regulations and policies faced by the project are minimal. The current new energy policies and financial subsidy policies are conducive to the development of the CBM industry. Second, there is no conflict between mining rights and air rights in the block where the project is located. In addition, the major licensing and approval procedures for the project are also complete. Third, the coal reservoirs in the block where the project is located have

good porosity, high permeability, high gas content and gas saturation, which are conducive to the development of CBM. Finally, the economic operational risks of the project are also small. The project's return on investment and CBM production are at a high level.

(8) In summary, the overall risk of the project is small and within an acceptable range, indicating that the project has development value. The assessment results are in good agreement with the actual operation of the project, indicating that the established risk assessment model is feasible and effective.

5. Conclusions

With the development of the CBM industry in various countries around the world, a mature and complete risk assessment system is needed to accurately determine the overall risk level and local risk weakness of CBM development projects. This paper provides a reference for government policy formulation, investor project feasibility analyses, insurance company insurance premium rate determination, and project operator risk perception improvement ability. This study fills the gap in the literature concerning the above requirements. The main contributions of this paper are as follows:

(1) Taking a CBM development project as the research object and considering the risk characteristics of exploration, mining, gathering and market application involved in the project life cycle, a CBM development risk assessment index system consisting of six first grade indexes and 45 second grade indexes is constructed.

(2) Based on the SEWM, the weights of the risk assessment indexes are calculated. Then, the MEEM is used to construct the theoretical model of CBM development risk assessment.

(3) A case study of a CBM development project in the southern Qinshui Basin of China was carried out using the established CBM development risk assessment model. The results show that the overall risk level of the CBM development project is Grade II, indicating that the overall risk of the project is small and within an acceptable range, although the local risk of the project needs to be rectified in a timely fashion. The assessment results are in good agreement with the actual operation of the project, indicating that the established risk assessment model has good applicability and effectiveness.

(4) The study makes three contributions. First, the research results can help relevant government departments formulate policies to reduce the risks faced by the CBM industry. Second, this work can provide a reference for investors to evaluate the feasibility of CBM development projects and for insurance companies to determine the insurance rate of CBM development projects. Finally, the research results are also conducive to improving the risk awareness and risk perception of CBM project operators. Therefore, the accuracy of managers' risk aversion decisions can be improved, and the waste of resources and property loss caused by the failure of risk management and control in the daily production process of the project can be reduced.

Author Contributions: Data curation, W.W.; Formal analysis, W.W., S.L. and Y.Z.; Methodology, W.W. and Y.Z.; Supervision, S.L.; Writing—original draft, W.W.; Writing—review and editing, Y.Z. and S.M.

Funding: This research was funded by [National Natural Science Foundation of China], grant number [51474151], and [PICC disaster research fund], grant number [D14-01].

Conflicts of Interest: The authors declare no conflicts of interest.

Abbreviations

CBM   Coalbed methane
SEWM   Structure entropy weight method
MEEM   Matter-element extension method

Appendix A

Table A1. Index analysis.

First Grade Indexes (r_i)	Index Analysis	Second Grade Indexes (r_{ik})	Index Analysis	Reference
Laws, regulations and policies (r_1)	Such risks refer to the laws, regulatory constraints and policy controls that are affected during the development of CBM. They also include uncertainties, such as audit approvals for CBM development.	Mineral rights and gas rights conflict (r_{11})	In China, coal mines are licensed by the State Ministry of Land and Resources and the Provincial Department of Land and Resources, while CBM is licensed by the Ministry of Land and Resources. However, there are often overlapping conflicts between CBM and coal at the level of hydrocarbon accumulation. In addition, China has not always made good distinctions between the block planning of coal mines and CBM. Therefore, in many blocks, there are both coal mining rights issued by the Provincial Department of Land and Resources and CBM mining rights granted by the State Ministry of Land and Resources. This type of conflict has caused many worries for investors.	Zhang et al. [29]
		Major licenses and approvals (r_{12})	CBM development also involves a number of major licensing and approval risks. Investors and companies should strictly inspect the exploration and mining licenses, safety production licenses and dangerous chemicals business licenses for CBM development projects. The absence of any of the above documents poses a significant risk to the operation of the project.	PRC [38]
		Foreign cooperation franchise (r_{13})	Foreign cooperation can make up for the shortage of funds for CBM projects and introduce advanced technology and management experience, thus promoting the development of CBM projects.	PRC [39]
		Government financial subsidies (r_{14})	Increasing financial subsidies can increase corporate liquidity, reduce production costs, and ease corporate investment pressures. Lowering or eliminating financial subsidies will increase competition in the CBM market and restrict the initial development of CBM enterprises.	Zhang et al. [13]
		Resource tax reform (r_{15})	Increasing the resource tax will increase the production cost of CBM enterprises, reduce the profit margin, and hinder the enthusiasm of enterprises. In contrast, reducing or cancelling unreasonable tax and fee systems and increasing tax incentives will incentivize CBM enterprises.	Zhang et al. [13]
		New energy policy (r_{16})	In the short run, CBM resources account for a certain proportion o' China's energy consumption. However, in the long run, with the development of new energy technologies, there will be inevitable changes i' China's energy structure, affecting the supply and demand of CBM. Therefore, the implementation of the national new energy policy may introduce risks to CBM development.	PRC [39]

Table A1. *Cont.*

First Grade Indexes (r_i)	Index Analysis	Second Grade Indexes (r_{ik})	Index Analysis	Reference
Resource characteristics (r_2)	Risks associated with resource characteristics mainly come from the CBM resources themselves and geological factors. Such risks refer to factors that are unstable or not widely available, due to resource conditions, resource distribution, CBM quality and geological factors.	Gas content (r_{21})	The gas content of CBM determines the critical desorption pressure. The higher the critical desorption pressure, the smaller the magnitude of the pressure reduction required for the CBM well. This also means that the earlier the CBM well starts to produce gas, the more methane gas can be desorbed by the coal seam. Therefore, insufficient CBM resources may pose risks to CBM development.	Roadifer et al. [10]
		Permeability (r_{22})	Permeability is the most critical factor in controlling the flow of CBM in coal reservoirs and the gas production of gas wells. It determines whether CBM can be successfully recovered. Therefore, permeability directly affects the economic benefits of CBM development, and low permeability of the coal seam may pose risks to the development of CBM.	Zhang et al. [13]
		Reservoir pressure (r_{23})	In the production process of CBM, a reasonable grasp of the trend of bottom hole flow pressure can reduce reservoir sensitivity, and thus, increase the production of CBM. Therefore, the size and distribution of coal reservoir pressure affect not only the enrichment of CBM, but also the production of CBM.	Zhang et al. [13]
		Porosity (r_{24})	Porosity is a key factor in determining the adsorption, permeability and strength properties of coal. By measuring the porosity and pressure of the CBM, the content of free CBM in the coal can be obtained. As a transport channel connecting the adsorption volume with the free surface, the pores also constitute the adsorption, diffusion and permeation system of CBM. Therefore, poor coal seam porosity may pose risks to CBM development.	Mares et al. [16]
		Hydrogeological conditions (r_{25})	The groundwater system controls the adsorption and accumulation of CBM through formation pressure. This gas control can cause the CBM to escape, and it can also conserve the CBM. In the development of CBM, CBM wells generally require a large amount of drainage and pressure reduction to produce gas. Therefore, regional hydrogeological conditions are the main factors determining the selection of well network, downhole equipment and dehydration process in the development of CBM. Therefore, unsuitable hydrogeological conditions may introduce risks to the development of CBM.	Lei et al. [31]
		Coal seam area (r_{26})	At the initial stage of the CBM development process, the coal seam area containing CBM is an important measure. In general, the larger the area of the gas-bearing coal seam, the higher the CBM content. Therefore, a poor coal seam area may pose risks to CBM development.	Zhai et al. [32]
		Coal seam thickness (r_{27})	In the case of a certain degree of gas saturation, the thickness of the coal seam directly determines the amount of gas-bearing resources of the gas reservoir. The thicker the coal seam, the more abundant the CBM that converges into the wellbore and the higher the CBM production. Therefore, insufficient coal seam thickness may introduce risks to CBM development.	Zhai et al. [32]
		Buried depth (r_{28})	The burial depth of coal seams is the main source of pressure. As the burial depth increases, the pressure will increase, which will increase the coalbed's ability to adsorb methane. In addition, if the geological structural conditions change so that the burial depth of the coal seam changes, the pressure will be changed, thus, affecting the preservation of the CBM reservoir. Therefore, the burial depth of CBM may introduce risks to the development of CBM.	Ge et al. [30]

Table A1. *Cont.*

First Grade Indexes (r_i)	Index Analysis	Second Grade Indexes (r_{ik})	Index Analysis	Reference
Engineering technology (r_3)	Engineering technology risks are mainly caused by inadaptability of geological evaluation, exploration and development, ground supporting engineering, dynamic monitoring and analysis and other technologies.	Geological evaluation technique (r_{31})	Geological evaluation technology plays a key role in understanding the geological conditions of CBM resources and the law of hydrocarbon accumulation. Advanced geological resource evaluation technology can accurately detect the distribution and enrichment of CBM resources. Therefore, backward geological evaluation technology will introduce risks to the development of CBM.	Mu et al. [17]
		Extraction technology (r_{32})	The extraction technology involves the adaptability of coal seams, which will directly affect the recovery and production of CBM wells. Moreover, the extraction technology and the drainage system should be coordinated; otherwise, reservoir ground stress and effective stress sensitivity damage will be caused, and the permeability will decrease. Therefore, backward mining technology may introduce risks to the development of CBM.	Jie et al. [33]
		Development process technology (r_{33})	The technology used in CBM development, the means for increasing production and the effect of increasing production vary across projects. At present, China's CBM development process technology mainly uses fracturing stimulation technology, and the fracturing effect directly affects the production of CBM wells. Therefore, if the CBM development process is backward, it will introduce risks to the development of CBM.	Jie et al. [33]
		Gas gathering technology (r_{34})	Gas gathering technology can effectively reduce the pressure of the gas production pipeline and the operating cost of the pipeline network and improve the operating efficiency and the development of CBM. Therefore, backward gas gathering process technology will increase operating costs and reduce operating efficiency, which will introduce risks to the development of CBM.	Jie et al. [33]
		Treatment technology (r_{35})	The treatment technology can not only increase the methane content in the CBM, but also realize the compression or liquefaction of the CBM, thereby greatly increasing the storage and transportation volume of the CBM and driving the development of the CBM enterprises. Therefore, if the treatment technology is backward, the quality, reserves and transportation of CBM will be reduced, thus, introducing economic risks to CBM development.	Jie et al. [33]
		Transportation technology (r_{36})	At present, CBM is mainly transported by tank trucks and pipelines. Among them, the tank truck transportation risk is greater because it involves more human error, and the delivery volume is much smaller than that of pipeline transportation. Transportation by long-distance pipeline network can reduce the potential for explosions caused by low-concentration CBM. Pipeline transportation technology can not only reduce the purification cost, but also greatly increase the transportation capacity of CBM. Therefore, backward transportation technology will also affect the development of CBM enterprises.	Jie et al. [33]
		Dynamic monitoring and analysis (r_{37})	Dynamic monitoring and analysis technology can observe and monitor the dynamic production of CBM exploration and development activities. This technology can determine the pressure distribution, connectivity of gas wells, and utilization of coal reservoirs. The technology can also guide production adjustment, which is conducive to improving the description accuracy of CBM reservoirs and the development of coal seams, thereby increasing the recovery rate of CBM. Therefore, the backwardness of dynamic monitoring and analysis technology will introduce risks to the development of CBM.	Lyu et al. [34]
		Drainage system (r_{38})	Unreasonable drainage systems often cause reservoir stress sensitivity damage. An unreasonable drainage system may significantly harm reservoir permeability. Too much damage may make it difficult or impossible to return the reservoir to the original state and may result in the CBM well not producing gas at all. Therefore, an unreasonable CBM drainage system may introduce risks to CBM development.	Zhang et al. [13]

Table A1. *Cont.*

First Grade Indexes (r_i)	Index Analysis	Second Grade Indexes (r_{ik})	Index Analysis	Reference
Economic operation (r_4)	Such risks refer to economic risks caused by macroeconomic weakness, price fluctuations, and capital supply in actual economic activities. These factors are potential uncertainties in the economic base, economic situation, and the ability to solve economic problems in CBM development projects.	CBM price fluctuation (r_{41})	The price of CBM is a very sensitive factor that is greatly affected by the national energy development and resource policy adjustment and directly affects the economic benefits of CBM development. Therefore, fluctuations in the price of CBM will adversely affect the development of the CBM industry.	Luo et al. [15]
		Project cost (r_{42})	High cost and low profit are the status quo of CBM development, which shows that if the project operator does not have a certain capital base, it will not be able to withstand the risks of CBM development. The level of mining costs also determines the level of profit. In recent years, although China's CBM demand and prices have been rising, due to the increase in staff salaries and environmental compensation fees, the project mining costs have also increased, which has led to a growth rate of CBM development profit that is far less than the increase in CBM prices. Therefore, the project mining cost has become another important factor restricting the development of the CBM industry.	Luo et al. [15]
		Return on investment (r_{43})	For CBM development enterprises, operating profit is the driving force for survival and development. If this driving force is lost, the solvency of the company will decline. Enterprises will fall into the predicament of production and development. Therefore, a low return on investment may introduce risks to CBM development.	Xia et al. [35]
		CBM production (r_{44})	CBM production stability directly affects the daily operation of the project, which means that if the project's CBM production is unstable or insufficient, it will bring high economic risks to the company.	Su et al. [19]
		Macroeconomics (r_{45})	CBM development is closely related to the growth of the national economy. When the national economy develops slowly and even stagnates, the demand and price of CBM will fall. In addition, the risks brought about by the macroeconomic downturn to CBM development are reflected in interest rates, money supply and inflation. Therefore, the macroeconomic downturn may introduce risks to CBM development.	Senthi et al. [11]
		Funds recovery (r_{46})	Some CBM development companies do not know enough about the credit rating of customers in the credit sales process and blindly sell to them, which will result in a large number of receivables that cannot be recovered for a long time until they become bad debts. On the other hand, in China's CBM development of current assets, the proportion of inventory is relatively large, and many enterprises are characterized by overstocking and inventory. Assets are occupied by debtors and stocks for a long time, making CBM development lack sufficient liquidity to reinvest or return debts due. The above situations will seriously affect the liquidity and safety of CBM development assets.	Xia et al. [35]
		External market resources (r_{47})	A typical feature of external market resources is market power, which refers to the influence and control of the market by economic entities. Market power refers to companies' ability to charge market prices to seek benefits and gain influence in the market and dominate stakeholder behavior. Market resources are expressed in various aspects, such as corporate image, goodwill and reputation. If the external market resources of the enterprise are poor, there may be risks to the development of CBM.	Zhang et al. [13]
		External network resources (r_{48})	External network resources mainly refer to the social relationships between enterprises and stakeholders. Many aspects of the CBM development process, such as access to funds, raw materials, production, sales promotion, and access to relevant policies, entail external relations. Relevant literature surveys have found that good external network resources can promote the development of a project. Conversely, the lack of such resources can introduce risks to CBM development.	Zhang et al. [13]
		Market demand (r_{49})	The market demand for clean energy is crucial to the development of the CBM industry. If the demand for clean energy is strong, it will be beneficial to the economic development of CBM development enterprises. If not, it will introduce risks to CBM development, due to the imbalance between supply and demand.	Chen et al. [12]

Table A1. *Cont.*

First Grade Indexes (r_i)	Index Analysis	Second Grade Indexes (r_{ik})	Index Analysis	Reference
Organizational management (r_5)	Organizational management not only provides the mechanism and framework for the development and operation of CBM, but also arranges personnel and resources for project operation. Organizational disruption and management failure are the least noticeable potential risks in the project's operations. It is also the root cause of the failure of the project management operational system.	Organizational structure adaptability (r_{51})	The organizational structure of the project determines not only how the strategic objectives and policies are established, but also the resource allocation efficiency of the project. The unreasonable organizational structure will increase the internal coordination cost of the enterprise, affect the efficiency and effectiveness of organizational decision-making, and reduce the efficiency of resource allocation. Therefore, an unreasonable organizational structure may introduce risks to the development of CBM.	Zhang et al. [13]
		Management coordination and communication (r_{52})	Due to the special nature of the workplace, the working surface needs to be constantly changing. Good coordination and communication within the organization not only ensure the smooth and efficient operation of the CBM development process, but also improve labor productivity and safe production management efficiency. Good coordination and communication skills can also maximize the overall efficiency of the project through effective systems, appropriate business complementarities, and good interpersonal relationships, thereby improving overall competitiveness. Conversely, if the company's management coordination and communication skills are insufficient, it may introduce risks to CBM development.	Zhang et al. [13]
		Process management (r_{53})	In the process of production and operation, the CBM development project involves many departments, and the production process is very complicated. The project's process management involves coordination between departments, reduction of operating costs and a guarantee of safe production. Therefore, inadequate process management may pose risks to CBM development.	Zhang et al. [13]
		Organizational management fineness (r_{54})	The main embodiment of the refined management of the CBM development process is the refined production method. It determines how to improve technology, eliminate waste, increase recovery rates, and efficiently use limited resources. Refined production methods are also an effective way to increase resources. Therefore, the lack of refined management of the CBM development process will introduce risks to CBM development.	Zhang et al. [13]
		Resource allocation capability (r_{55})	Resource allocation capability refers to maximizing the role of the resources a company owns and promoting the achievement of development goals. For CBM development projects, internal resources, such as CBM resources, financial resources and human resources, as well as external resources closely related to CBM development, play a key role in industrial development. How to use these resources reasonably and to their full potential without causing waste is an important issue that must be dealt with in the development of CBM. Therefore, an inefficient resource allocation capacity may introduce risks to CBM development.	Zhang et al. [13]

Table A1. *Cont.*

First Grade Indexes (r_i)	Index Analysis	Second Grade Indexes (r_{ik})	Index Analysis	Reference
Safety and emergency protection (r_6)	Safety and emergency protection risks affect the entire process of CBM development. The CBM development process involves a wide range of content, many links, complicated operating procedures, and a large safety supervision span. Any problem with any link can introduce risks to a CBM development project.	Safety technology and equipment (r_{61})	In the process of CBM development, relevant technologies and equipment are required for effective safety protection. Compared with that in other countries, there is still a gap in the safety technology of CBM development in China, and the research and development of safety equipment are still far behind. Therefore, CBM development safety accidents have occasionally occurred in China. Therefore, the backwardness of safety technology and equipment will introduce risks to CBM development.	Kirchgessne et al. [18]
		Hidden danger investigation and treatment (r_{62})	CBM is a flammable and explosive gas. In the operation of CBM projects, enterprises should attach great importance to the investigation of accident safety hazards. If hidden dangers are found in the inspection, they should be quickly rectified to avoid fire or explosion accidents. Therefore, if the company does not establish an effective long-term mechanism for potential hazard investigation and treatment, it will increase the risk of safe production of CBM development.	PRC [40]
		Safety training and education (r_{63})	Accidents are often caused by the unsafe behavior of people and the unsafe state of things. To reduce human error in the development of CBM and enhance employees' safety awareness, enterprises should organize training and education on a regular basis. Otherwise, safety production accidents caused by human error will cause casualties and halt production.	PRC [40]
		Safety culture (r_{64})	The construction of corporate safety culture is an effective guarantee for the safe production of the project. The system is rigid, and consciousness and culture are flexible. Only by orienting both aspects of safety can employees effectively prevent the occurrence of security incidents, due to their understanding and habits. Therefore, the lack of a corporate safety culture may introduce risks to CBM development.	Henriqson et al. [36]
		Safety investment (r_{65})	Safety investment affects the advancement of safety technology and equipment and the preparation of emergency supplies. Enterprises should maximize their safety investment funds, purchase advanced safety equipment and sufficient emergency materials, and regularly organize experts to train employees. If the company's safety investment is too small, it will indirectly affect other safety factors, thus, introducing risks to CBM development.	Zhao et al. [37]
		Emergency plan (r_{66})	China's Safe Production Law stipulates that the CBM production industry should formulate special emergency plans for all kinds of accidents and all hazards that may occur and have clear rescue procedures and specific emergency rescue measures. If the company does not formulate an emergency plan, accidents can easily spread, resulting in more serious casualties and property losses.	PRC [40]
		Emergency drill frequency (r_{67})	Conducting emergency drills can leave a deeper impression on participants by simulating the emergency response process of CBM development accidents. Let employees truly understand accidents intuitively and emotionally and improve their vigilance against accident risk sources. Emergency drills not only enhance employees' emergency awareness and accident handling skills, but also improve their self-help and mutual rescue capabilities. Therefore, insufficient frequency of emergency drills will introduce risks to CBM development.	PRC [41]
		Emergency supplies reserve (r_{68})	In the event of a safety production accident, if emergency supplies cannot be provided in a timely and comprehensive manner, it is difficult to achieve accident rescue. Therefore, insufficient emergency material reserves will introduce risks to CBM development.	PRC [41]
		Emergency rescue team (r_{69})	China's Hazardous Chemicals Safety Management Regulations clearly state that the CBM development industry should establish an emergency rescue team consisting of full-time or part-time personnel. In addition, emergency rescue teams must organize emergency rescue operations in the event of an accident and assist in the investigation of risk hidden dangers during normal project operation. Therefore, if the companies do not establish emergency rescue teams, they will introduce risks to CBM development.	PRC [41]

Appendix B

Table A2. Weight distribution of the second grade indexes of r_1.

	Index r_{11}	Index r_{12}	Index r_{13}	Index r_{14}	Index r_{15}	Index r_{16}
Group A	1	2	6	3	5	4
Group B	2	1	5	4	3	6
Group C	2	1	5	3	4	4
Group D	2	1	4	3	6	5
b_j	0.941	0.980	0.549	0.798	0.615	0.586
$b_j max$	1.000	1.000	0.712	0.827	0.827	0.712
$b_j max - b_j$	0.059	0.020	0.163	0.029	0.212	0.126
$b_j min$	0.921	0.921	0.356	0.712	0.356	0.356
$b_j - b_j min$	0.020	0.059	0.193	0.086	0.259	0.230
σ_j	0.040	0.040	0.178	0.058	0.236	0.178
$1-\sigma_j$	0.961	0.961	0.822	0.943	0.765	0.822
X_j	0.904	0.941	0.451	0.752	0.470	0.482
Weight	0.226	0.235	0.113	0.188	0.118	0.120

The calculated membership matrix B_1 is based on Formula (4), and m is set as 8.

$$B_1 = \begin{bmatrix} 1.000 & 0.921 & 0.356 & 0.827 & 0.565 & 0.712 \\ 0.921 & 1.000 & 0.565 & 0.712 & 0.827 & 0.356 \\ 0.921 & 1.000 & 0.565 & 0.827 & 0.712 & 0.712 \\ 0.921 & 1.000 & 0.712 & 0.827 & 0.356 & 0.565 \end{bmatrix}$$

Table A3. Weight distribution of the second grade indexes of r_2.

	Index r_{21}	Index r_{22}	Index r_{23}	Index r_{24}	Index r_{25}	Index r_{26}	Index r_{27}	Index r_{28}
Group A	2	1	3	4	5	7	7	6
Group B	3	1	2	5			8	6
Group C	3	2	1	6	4	8	7	5
Group D	1	2	3	4	6	7	8	5
b_j	0.929	0.973	0.929	0.749	0.749	0.454	0.408	0.682
$b_j max$	1.000	1.000	1.000	0.815	0.815	0.500	0.500	0.732
$b_j max - b_j$	0.071	0.027	0.071	0.067	0.067	0.046	0.092	0.051
$b_j min$	0.886	0.946	0.886	0.631	0.631	0.315	0.315	0.631
$b_j - b_j min$	0.044	0.027	0.044	0.118	0.118	0.138	0.092	0.051
σ_j	0.057	0.027	0.057	0.092	0.092	0.092	0.092	0.051
$1-\sigma_j$	0.943	0.973	0.943	0.908	0.908	0.908	0.908	0.949
X_j	0.876	0.947	0.876	0.680	0.680	0.412	0.370	0.647
Weight	0.160	0.173	0.160	0.124	0.124	0.075	0.067	0.118

The calculated membership matrix B_2 is based on Formula (4), and m is set as 10.

$$B_2 = \begin{bmatrix} 0.946 & 1.000 & 0.886 & 0.815 & 0.732 & 0.500 & 0.500 & 0.631 \\ 0.886 & 1.000 & 0.946 & 0.732 & 0.815 & 0.500 & 0.315 & 0.631 \\ 0.886 & 0.946 & 1.000 & 0.631 & 0.815 & 0.315 & 0.500 & 0.732 \\ 1.000 & 0.946 & 0.886 & 0.815 & 0.631 & 0.500 & 0.315 & 0.732 \end{bmatrix}$$

Table A4. Weight distribution of the second grade indexes of r_3.

	Index r_{31}	Index r_{32}	Index r_{33}	Index r_{34}	Index r_{35}	Index r_{36}	Index r_{37}	Index r_{38}
Group A	1	2	4	5	3	7	6	4
Group B	1	3	2	6	4	8	7	5
Group C	2	1	3	4	7	6	5	3
Group D	1	2	5	4	3	7	6	3
b_j	0.987	0.945	0.845	0.749	0.772	0.487	0.624	0.830
$b_j max$	1.000	1.000	0.946	0.815	0.886	0.631	0.732	0.886
$b_j max - b_j$	0.013	0.055	0.101	0.067	0.114	0.144	0.109	0.056
$b_j min$	0.946	0.886	0.732	0.631	0.500	0.315	0.500	0.732
$b_j - b_j min$	0.040	0.059	0.113	0.118	0.272	0.171	0.124	0.097
σ_j	0.027	0.057	0.107	0.092	0.193	0.158	0.116	0.077
$1 - \sigma_j$	0.972	0.943	0.893	0.908	0.807	0.842	0.884	0.923
X_j	0.960	0.891	0.755	0.680	0.623	0.410	0.551	0.766
Weight	0.170	0.158	0.134	0.121	0.111	0.073	0.098	0.136

The calculated membership matrix B_3 is based on the Formula (4), and m is set as 10.

$$B_3 = \begin{bmatrix} 1.000 & 0.946 & 0.815 & 0.732 & 0.886 & 0.500 & 0.631 & 0.815 \\ 1.000 & 0.886 & 0.946 & 0.631 & 0.815 & 0.315 & 0.500 & 0.732 \\ 0.946 & 1.000 & 0.886 & 0.815 & 0.500 & 0.631 & 0.732 & 0.886 \\ 1.000 & 0.946 & 0.732 & 0.815 & 0.886 & 0.500 & 0.631 & 0.886 \end{bmatrix}$$

Table A5. Weight distribution of the second grade indexes of r_4.

	Index r_{41}	Index r_{42}	Index r_{43}	Index r_{44}	Index r_{45}	Index r_{46}	Index r_{47}	Index r_{48}	Index r_{49}
Group A	8	2	4	3	7	5	6	9	1
Group B	6	1	2	3	8	7	4	5	1
Group C	5	1	3	2	4	6	7	8	2
Group D	7	3	4	2	6	5	8	9	1
b_j	0.639	0.964	0.887	0.929	0.656	0.714	0.656	0.464	0.989
$b_j max$	0.778	1.000	0.954	0.954	0.845	0.778	0.845	0.778	1.000
$b_j max - b_j$	0.139	0.036	0.067	0.0256	0.189	0.064	0.189	0.314	0.011
$b_j min$	0.477	0.903	0.845	0.903	0.477	0.602	0.477	0.301	0.954
$b_j - b_j min$	0.162	0.061	0.042	0.026	0.179	0.112	0.179	0.163	0.034
σ_j	0.151	0.048	0.055	0.026	0.184	0.088	0.184	0.239	0.023
$1 - \sigma_j$	0.850	0.952	0.945	0.974	0.816	0.912	0.816	0.761	0.978
X_j	0.543	0.918	0.838	0.905	0.535	0.651	0.535	0.354	0.966
Weight	0.087	0.147	0.134	0.145	0.086	0.104	0.086	0.057	0.155

The calculated membership matrix B_4 is based on Formula (4), and m is set as 11.

$$B_4 = \begin{bmatrix} 0.477 & 0.954 & 0.845 & 0.903 & 0.602 & 0.778 & 0.699 & 0.301 & 1.000 \\ 0.699 & 1.000 & 0.954 & 0.903 & 0.477 & 0.602 & 0.845 & 0.778 & 1.000 \\ 0.778 & 1.000 & 0.903 & 0.954 & 0.845 & 0.699 & 0.602 & 0.477 & 0.954 \\ 0.602 & 0.903 & 0.845 & 0.954 & 0.699 & 0.778 & 0.477 & 0.301 & 1.000 \end{bmatrix}$$

Table A6. Weight distribution of the second grade indexes of r_5.

	Index r_{51}	Index r_{52}	Index r_{53}	Index r_{54}	Index r_{55}
Group A	1	3	4	5	2
Group B	1	2	3	4	2
Group C	2	4	1	5	3
Group D	1	4	2	5	3
b_j	0.975	0.725	0.821	0.443	0.836
$b_j max$	1.000	0.898	0.898	0.613	0.898
$b_j max- b_j$	0.025	0.174	0.077	0.170	0.062
$b_j min$	0.898	0.613	0.613	0.387	0.774
$b_j- b_j min$	0.076	0.111	0.208	0.057	0.062
σ_j	0.051	0.143	0.143	0.113	0.062
$1\text{-}\sigma_j$	0.949	0.857	0.857	0.887	0.938
X_j	0.925	0.621	0.704	0.393	0.784
Weight	0.270	0.181	0.205	0.115	0.229

The calculated membership matrix B_5 is based on Formula (4), and m is set as 7.

$$B_5 = \begin{bmatrix} 1.000 & 0.774 & 0.613 & 0.387 & 0.898 \\ 1.000 & 0.898 & 0.774 & 0.613 & 0.898 \\ 0.898 & 0.613 & 1.000 & 0.387 & 0.774 \\ 1.000 & 0.613 & 0.898 & 0.387 & 0.774 \end{bmatrix}$$

Table A7. Weight distribution of the second grade indexes of r_5.

	Index r_{61}	Index r_{62}	Index r_{63}	Index r_{64}	Index r_{65}	Index r_{66}	Index r_{67}	Index r_{68}	Index r_{69}
Group A	1	2	4	8	3	5	6	7	7
Group B	2	3	4	8	1	5	5	7	6
Group C	1	4	2	9	3	7	8	6	5
Group D	1	3	2	8	1	7	6	5	4
b_j	0.989	0.901	0.900	0.433	0.952	0.690	0.663	0.670	0.731
$b_j max$	1.000	0.954	0.954	0.477	1.000	0.778	0.778	0.778	0.845
$b_j max- b_j$	0.011	0.053	0.055	0.044	0.048	0.088	0.115	0.108	0.114
$b_j min$	0.954	0.845	0.845	0.301	0.903	0.602	0.477	0.602	0.602
$b_j- b_j min$	0.034	0.056	0.055	0.132	0.048	0.088	0.186	0.068	0.129
σ_j	0.023	0.055	0.055	0.088	0.048	0.088	0.151	0.088	0.122
$1\text{-}\sigma_j$	0.977	0.945	0.945	0.912	0.952	0.912	0.850	0.912	0.878
X_j	0.966	0.852	0.851	0.395	0.905	0.629	0.563	0.611	0.642
Weight	0.151	0.133	0.133	0.062	0.141	0.098	0.088	0.0953	0.100

The calculated membership matrix B_6 is based on Formula (4), and m is set as 11.

$$B_6 = \begin{bmatrix} 1.000 & 0.954 & 0.845 & 0.477 & 0.903 & 0.778 & 0.699 & 0.602 & 0.602 \\ 0.954 & 0.953 & 0.845 & 0.477 & 1.000 & 0.778 & 0.778 & 0.602 & 0.699 \\ 1.000 & 0.845 & 0.954 & 0.301 & 0.903 & 0.602 & 0.477 & 0.699 & 0.778 \\ 1.000 & 0.903 & 0.945 & 0.477 & 1.000 & 0.602 & 0.699 & 0.778 & 0.845 \end{bmatrix}$$

Appendix C

(1) The correlation degree matrix of the first grade index r_2 is calculated as follows:

$$K(r_2) = (\omega_{21}, \omega_{22}, \omega_{23}, \omega_{24}, \omega_{25}, \omega_{26}, \omega_{27}, \omega_{28})(k_j(r_2)) =$$

$$(0.160, 0.173, 0.160, 0.124, 0.124, 0.075, 0.067, 0.118) \begin{bmatrix} -0.222 & 0.400 & -0.300 & -0.580 & -0.720 \\ -0.143 & 0.200 & -0.400 & -0.640 & -0.760 \\ 0.571 & -0.267 & -0.633 & -0.780 & -0.853 \\ -0.273 & -0.333 & -0.200 & -0.520 & -0.680 \\ 0.077 & -0.067 & -0.533 & -0.720 & -0.813 \\ -0.297 & 0.182 & -0.133 & -0.480 & -0.653 \\ -0.174 & 0.267 & -0.367 & -0.620 & -0.747 \\ 0.182 & -0.133 & -0.567 & -0.740 & -0.826 \end{bmatrix}$$

$$= (-0.006, 0.105, -0.411, -0.647, -0.765)$$

(2) The correlation degree matrix of the first grade index r_3 is calculated as follows:

$$K(r_3) = (\omega_{31}, \omega_{32}, \omega_{33}, \omega_{34}, \omega_{35}, \omega_{36}, \omega_{37}, \omega_{38})(k_j(r_3)) =$$

$$(0.170, 0.158, 0.134, 0.121, 0.111, 0.073, 0.098, 0.136) \begin{bmatrix} -0.770 & -0.067 & -0.533 & -0.720 & -0.813 \\ -0.297 & 0.182 & -0.133 & -0.480 & -0.653 \\ -0.258 & 0.438 & -0.233 & -0.540 & -0.693 \\ -0.143 & 0.200 & -0.400 & -0.640 & -0.760 \\ -0.174 & 0.267 & -0.367 & -0.620 & -0.747 \\ -0.347 & -0.059 & 0.067 & -0.360 & -0.573 \\ -0.241 & 0.467 & -0.267 & -0.560 & -0.707 \\ -0.286 & 0.250 & -0.167 & -0.500 & -0.667 \end{bmatrix}$$

$$= (-0.075, 0.205, -0.276, -0.566, -0.711);$$

(3) Correlation degree matrix of the first grade index r_4 is calculated as follows:

$$K(r_4) = (\omega_{41}, \omega_{42}, \omega_{43}, \omega_{44}, \omega_{45}, \omega_{46}, \omega_{47}, \omega_{48}, \omega_{49})(k_j(r_4)) =$$

$$(0.087, 0.147, 0.134, 0.145, 0.086, 0.104, 0.086, 0.057, 0.155) \begin{bmatrix} -0.105 & 0.133 & -0.433 & -0.660 & -0.773 \\ -0.241 & 0.467 & -0.267 & -0.560 & -0.707 \\ -0.273 & 0.333 & -0.200 & -0.520 & -0.680 \\ -0.317 & 0.077 & -0.067 & -0.440 & -0.627 \\ -0.059 & 0.067 & -0.467 & -0.820 & -0.747 \\ 0.077 & -0.067 & -0.533 & -0.720 & -0.813 \\ -0.397 & -0.214 & 0.158 & -0.120 & -0.413 \\ -0.143 & 0.200 & -0.400 & -0.640 & -0.760 \\ -0.241 & 0.467 & -0.267 & -0.460 & -0.707 \end{bmatrix}$$

$$= (-0.204, 0.200, -0.260, -0.540, -0.694);$$

(4) Correlation degree matrix of the first grade index r_5 is calculated as follows:

$$K(r_5) = (\omega_{51}, \omega_{52}, \omega_{53}, \omega_{54}, \omega_{55})(k_j(r_5)) =$$

$$(0.270, 0.181, 0.250, 0.115, 0.229) \begin{bmatrix} 0.000 & 0.000 & -0.500 & -0.700 & -0.800 \\ -0.222 & 0.400 & -0.300 & -0.580 & -0.720 \\ -0.308 & 0.125 & -0.100 & -0.460 & -0.640 \\ -0.142 & 0.200 & -0.400 & -0.640 & -0.760 \\ -0.297 & 0.182 & -0.133 & -0.480 & -0.653 \end{bmatrix}$$

$$= (-0.188, 0.163, -0.286, -0.572, -0.714);$$

(5) Correlation degree matrix of the first grade index r_6 is calculated as follows:

$$K(r_6) = (\omega_{61}, \omega_{62}, \omega_{63}, \omega_{64}, \omega_{65}, \omega_{66}, \omega_{67}, \omega_{68}, \omega_{69})(k_j(r_6)) =$$

$$(0.151, 0.133, 0.133, 0.062, 0.141, 0.098, 0.088, 0.095, 0.100) \begin{bmatrix} -0.286 & 0.250 & -0.167 & -0.500 & -0.667 \\ -0.241 & 0.467 & -0.267 & -0.560 & -0.707 \\ -0.397 & -0.241 & 0.158 & -0.120 & -0.413 \\ -0.317 & 0.077 & -0.067 & -0.440 & -0.627 \\ -0.143 & -0.200 & 0.400 & -0.640 & -0.760 \\ -0.297 & 0.182 & -0.133 & -0.480 & -0.653 \\ -0.403 & -0.258 & 0.095 & -0.080 & -0.387 \\ -0.373 & -0.159 & 0.233 & -0.260 & -0.507 \\ -0.273 & 0.333 & -0.200 & -0.520 & -0.680 \end{bmatrix}$$

$$= (-0.295, 0.114, -0.103, -0.414, -0.610)$$

References

1. Zhou, C.; Huang, G.; Chen, J. A multi-objective energy and environmental systems planning model: management of uncertainties and risks for Shanxi province, China. *Energies* **2018**, *11*, 2723. [CrossRef]
2. Cheng, L.; Ge, Z.; Chen, J.; Ding, H.; Zou, L.; Li, K. A Sequential approach for integrated coal and gas mining of closely-spaced outburst coal seams: Results from a case study including mine safety improvements and greenhouse gas reductions. *Energies* **2018**, *11*, 3023. [CrossRef]
3. Wang, J.F.; Li, W.J. *China's Coal Mine Accidents and Comments of Safety Specialists*; China Coal Industry Publishing House: Beijing, China, 2002; pp. 1–8.
4. Zhang, J.J.; Xu, K.L.; You, G.; Wang, B.B.; Zhao, L. Causation analysis of risk coupling of gas explosion accident in Chinese underground coal mines. *Risk Anal.* **2019**, *4*, 1–13. [CrossRef] [PubMed]

5. Zhang, J.J.; David, C.; Xu, K.L.; You, G. Focusing on the patterns and characteristics of extraordinarily severe gas explosion accidents in Chinese coal mines. *Process Saf. Environ. Prot.* **2018**, *117*, 390–398. [CrossRef]

6. Knez, D.; Wiśniowski, R.; Owusu, W.A. Turning filling material into proppant for coalbed methane in Poland—Crush test results. *Energies* **2019**, *12*, 1820. [CrossRef]

7. Li, G.F.; Li, G.H.; Liu, G. Analysis on the ground extraction effect of coal-bed methane at typical area in Jincheng, China. *J. China Coal Soc.* **2014**, *39*, 1932–1937.

8. Karacan, C.O.; Ruiz, F.A.; Cote, M.; Phipps, S. Coal mine methane: A review of capture and utilization practices with benefits to mining safety and to greenhouse gas reduction. *Int. J. Coal Geol.* **2011**, *86*, 121–156. [CrossRef]

9. Liao, Y.Y.; Luo, D.K.; Li, W.D. Development strategy analysis of China's CBM. *Acta Pet. Sin.* **2012**, *33*, 1098–1102.

10. Roadifer, R.D. Coalbed methane parametric study: What's really important to production and when? In Proceedings of the SPE Annual Technical Conference and Exhibition, Denver, CO, USA, 5–8 October 2003. [CrossRef]

11. Senthil, B. Economic Feasibility of Coalbed Methane Projects Using Monte-Carlo and Hypercube Simulation. Ph.D. Thesis, Mississippi State University, Starkville, MS, USA, 1994.

12. Chen, Y.H.; Yang, Y.G.; Luo, J.H. Uncertainty Analysis of Coalbed Methane Economic Assessment with Montecarlo Method. *Procedia Environ. Sci.* **2012**, *12*, 640–645. [CrossRef]

13. Zhang, Y.C.; Yang, Y.G.; Luo, J.H. Study on risk assessment of CBM development based on optimized combination weighting. *J. Saf. Sci. Technol.* **2016**, *12*, 91–97.

14. Acquah-Andoh, E.; Putra, H.A.; Ifelebuegu, A.O.; Owusu, A. Coalbed methane development in Indonesia: Design and economic analysis of upstream petroleum fiscal policy. *Energy Policy* **2019**, *131*, 155–167. [CrossRef]

15. Luo, D.K.; Dai, Y.J. Economic evaluation of coalbed methane production in China. *Energy Policy* **2009**, *37*, 3883–3889. [CrossRef]

16. Mares, T.E.; Moore, T.A.; Moore, C.R. Uncertainty of gas saturation estimates in a subbituminous coal seam. *Int. J. Coal Geol.* **2009**, *77*, 320–327. [CrossRef]

17. Mu, F.Y.; Wang, H.Y.; Wu, J.T.; Sun, B. Practice of and suggestions on CBM development in China. *Nat. Gas Ind.* **2018**, *38*, 61–66.

18. Kirchgessne, D.A.; Masemore, S.S.; Piccot, S.D. Engineering and economic evaluation of gas recovery and utilization technologies at selected US mines. *Environ. Sci. Policy* **2002**, *5*, 397–409. [CrossRef]

19. Su, J.; Zhang, J.; Zhu, W. Improved methodology of economic evaluation of coalbed methane based on discounted cash flow analysis. *J. China Univ. Min. Technol.* **2018**, *47*, 176–183.

20. Lau, C.K.; Lai, K.K.; Lee, Y.P.; Du, J.Z. Fire risk assessment with scoring system, using the support vector machine approach. *Fire Saf. J.* **2015**, *78*, 188–195. [CrossRef]

21. Cheng, Q.Y. Structure entropy weight method to confirm the weight of evaluating index. *Syst. Eng. Theory Pract.* **2010**, *30*, 1225–1228.

22. Cai, W. Introduction of Extenics. *Syst. Eng. Theory Pract.* **1998**, *1*, 76–84.

23. Jin, Y.Q.; Hu, M.; Wu, Q.; Fang, C.X. Quantitative risk assessment model for high pressure oil and gas well control based on the matter-element extension. *J. Saf. Environ.* **2014**, *14*, 147–150.

24. Wang, J.M.; Jia, M.T.; Wang, J.; Chen, J.H.; Long, K.M. Risk evaluation of break-dam in tailings reservoir based on matter-element extension model. *J. Saf. Sci. Technol.* **2014**, *4*, 96–102.

25. Xie, Z.X.; Li, Y.Y.; Wu, X.R.; Wang, F.; Wang, W. Evaluation of the archive fire risk based on the matter-element and multi-level extension model. *J. Saf. Environ.* **2019**, *19*, 27–34.

26. Iribarren, D.; Martíngamboa, M.; O'Mahony, T.; Dufour, J. Screening of socio-economic indicators for sustainability assessment: A combined life cycle assessment and data envelopment analysis approach. *Int. J. Life Cycle Assess.* **2016**, *21*, 202–214. [CrossRef]

27. Elhami, B.; Raini, M.G.N.; Soheili-Fard, F. Energy and environmental indices through life cycle assessment of raisin production: A case study (Kohgiluyeh and Boyer-Ahmad Province, Iran). *Renew. Energy* **2019**, *141*, 507–515. [CrossRef]

28. Alvarez-Rodriguez, C.; Martin-Gamboa, M.; Iribarren, D. Sustainability-oriented management of retail stores through the combination of life cycle assessment and dynamic data envelopment analysis. *Sci. Total Environ.* **2019**, *68*, 49–60. [CrossRef]

29. Zhang, Y.S.; Niu, C.H. Research on risk factors and strategies of China's CBM industry development under the background of energy revolution. *Sci. Manag. Res.* **2019**, *37*, 68–73.
30. Ge, X.; Liu, D.; Cai, Y.; Wang, Y. Gas content evaluation of coalbed methane reservoir in the Fukang area of Southern Junggar Basin, northwest China by multiple geophysical logging methods. *Energies* **2018**, *11*, 1867. [CrossRef]
31. Lei, C.L.; Liu, X. Assessment of economic benefit for coalbed methane development project from geological hazard risk. *Coal Geol. Explor.* **2016**, *44*, 75–79.
32. Zhai, G.M.; He, W.Y. Occurrence features and exploration orientation of coalbed methane gas in China. *Nat. Gas Ind.* **2010**, *30*, 1–3.
33. Jie, M.X.; Ge, X.D.; Peng, C.Y.; Xiong, X.Y.; Xia, F. Advances in the CBM exploration and development techniques and their developing trend in China. *Nat. Gas Ind.* **2011**, *31*, 63–65.
34. Lyu, Y.M.; Tang, D.Z.; Li, Z.P.; Shao, X.J.; Xu, H. Fitting and predicting models for coalbed methane wells dynamic productivity. *J. China Coal Soc.* **2011**, *36*, 1481–1485.
35. Xia, L.Y.; Luo, D.K.; Dai, Y.J. Risk evaluation methods of CBM development projects. *Nat. Gas Ind.* **2012**, *32*, 117–120.
36. Henriqson, E.; Schuler, B.; Van Winsen, R.; Dekker, S.W.A. The constitution and effects of safety culture as an object in the discourse of accident prevention: A Foucauldian approach. *Saf. Sci.* **2014**, *70*, 465–476. [CrossRef]
37. Zhao, B.F.; Zhang, C.; Jia, B.S.; Zhai, C.X.; Ren, H.Z.; Guo, J.W. DMIP-MCDM based method of evaluating safety input for coal enterprises. *China Saf. Sci. J.* **2017**, *27*, 127–131.
38. *People's Republic of China Mineral Resources Law*; PRC Standing Committee of the National People's Congress: Beijing, China, 1986.
39. *Coalbed Methane Industry Policy*; PRC National Energy Board: Beijing, China, 2013.
40. *Safety Production Law of the People's Republic of China*; PRC Standing Committee of the National People's Congress: Beijing, China, 2002.
41. *Hazardous Chemicals Safety Management Regulations*; PRC State Council: Beijing, China, 2002.
42. Dalkey, N.C. The Delphi method: An experimental study of group opinion. *Futures* **1969**, *1*, 408–426. [CrossRef]
43. Liu, F.; Zhao, S.Z.; Weng, M.C.; Liu, Y.Q. Fire risk assessment for large-scale commercial buildings based on structure entropy weight method. *Saf. Sci.* **2017**, *94*, 26–40. [CrossRef]
44. Cai, W. Extension theory and its application. *Chin. Sci. Bull.* **1999**, *7*, 673–682. [CrossRef]
45. Cai, W.; Yang, C.Y. Basic theory and methodology on Extenics. *Chin. Sci. Bull.* **2013**, *58*, 1190–1199.
46. Guan, X.J. Evaluation method on risk grade of tunnel collapse based on extension connection cloud model. *J. Saf. Sci. Technol.* **2018**, *14*, 186–192.
47. Wang, Y.; Liu, D.; Cai, Y.; Li, X. Variation of petrophysical properties and adsorption capacity in different rank coals: An experimental study of coals from the Junggar, Ordos and Qinshui Basins in China. *Energies* **2019**, *12*, 986. [CrossRef]
48. Qin, Y.; Liang, J.S.; Shen, J.; Liu, Y.H.; Wang, C.W. Gas logging shows and gas reservoir types in tight sandstones and shales from Southern Qinshui Basin. *J. China Coal Soc.* **2014**, *39*, 1559–1565.
49. Zhang, Z.; Qin, Y.; Fu, X.H. The favorable developing geological conditions for CBM multi-layer drainage in southern Qinshui basin. *J. China Univ. Min. Technol.* **2014**, *43*, 1019–1024.
50. Li, Q.; He, J.J.; Cao, J. Physical characteristics of coalbed methane reservoir in Heshun Area of Qinshui Basin. *Oil Geophys. Prospect.* **2013**, *48*, 734–739.

Article

Life Cycle Sustainability Assessment of the Spanish Electricity: Past, Present and Future Projections

Guillermo San Miguel * and María Cerrato

Department of Chemical and Environmental Engineering, ETSII, Grupo de Agroenergética, C/José Gutiérrez Abascal, 2, Universidad Politécnica de Madrid, 28006 Madrid, Spain; m.cerrato@alumnos.upm.es
* Correspondence: g.sanmiguel@upm.es

Received: 4 March 2020; Accepted: 10 April 2020; Published: 13 April 2020

Abstract: This paper provides an investigation into the sustainability of the electrical system in Spain. The analysis covers historic inventories of power generation, installed capacity and technology mix since 1990 and also contemplates four alternative projections for 2030 and 2050. The sustainability is evaluated using eight indicators that provide objective information about the environmental (climate change, fossil depletion, ozone layer depletion, terrestrial acidification, human toxicity and photochemical smog), economic (levelized cost of electricity) and socio-economic (direct employment) performance of the system. The results show an increase in the magnitude of the environmental impacts between 1990 and 2008, due to a growing power demand triggered by economic expansion. After 2008, the environmental performance improves due to the economic recession and the penetration of renewable energies. Overall, the cost of power generation remains rather stable as rising expenses generated by renewables are compensated by a progressive reduction in the cost of fossil technologies. Direct employment generation has been strongly stimulated by the upsurge in renewables that has taken place in Spain after 2008. Regarding future scenarios, the results evidence that the most ambitious projections in terms of renewable penetration perform best in terms of environmental performance, employment generation and reduced costs (€/MWh). The significance of these benefits was particularly clear in the 2050 scenario. In the long term, the scenario considering higher fossil fuel contributions (ST) performed worst in all sustainability indicators.

Keywords: LCA; Spain; renewables; electricity; sustainability; carbon footprint; employment; LCOE

---

1. Introduction

Electricity is regarded as a fundamental commodity in modern societies. The availability of this energy vector is inextricably associated with economic prosperity, social progress and human development [1]. It is in this spirit that access to electricity has been incorporated into the Sustainable Development Goals (SDGs) (see Goal 7: Affordable and clean energy) defined under the much-acclaimed United Nations (UN) Agenda 2030 [2].

However, the deployment, operation and decommissioning of the infrastructures required to provide this essential service may generate substantial impacts (both positive or negative) on the sustainability of the natural and human environment. For instance, the industrial and commercial activity associated with the life cycle of a power plant (construction, operation, extraction and processing of fuels, decommissioning) will surely contribute to economic growth and job creation [3,4]. These actions will also be responsible for the deterioration of the surrounding environment, the magnitude of which would depend primarily on the generation technologies employed and the overall demand. These detrimental effects would be observed in impact categories such as global warming, acidification, toxicity and consumption of natural resources, to name a few. The nature and extent of these types of impact largely depend on the type of generation technology and energy source utilized [5–7].

Most countries around the world are currently involved in a profound transformation of their electricity systems. Over the last decade, Spain has been adapting to the requirements set by the European Union (EU) under the Renewable Energy Directive 2009/28/EC [8] and also to the responsibilities agreed to in the ongoing EU climate action plans [9]. In this context, the targets set under the 2020 Climate And Energy Package have given way to the more ambitious 2030 Climate and Energy Framework whose objectives include a 40% cut in greenhouse gas emissions (from 1990 levels), a 32% share for renewable energy and a 32.5% improvement in energy efficiency. The long-term European strategy for this transition is gradually starting to come to light in documents like A Clean Planet for All [10].

This transition towards a reduction in the use of fossil fuels and the promotion of local renewable energy sources has environmental, economic and geostrategic roots. The main environmental driver is the need to reduce greenhouse gas emissions (GHG) so as to avoid irreversible changes in the Earth's climate [11]. The uncertainty associated with the volatility of fossil fuel prices and the benefits from the opportunities generated by a progressive reduction in the cost of renewable energies are the two main economic drivers of such plan. This is reinforced by political instability of major producers of fossil resources (e.g., Venezuela, Persian Gulf and Arabian Peninsula, etc.) and the strategic inconveniences of energy dependency [12].

These changes in the total output and configuration of the electricity systems determine not only their environmental sustainability but also their socio-economic and economic performance. All these aspects need to be evaluated in order to understand the true consequences that these changes may bring about, so that positive aspects may be maximized while negative impacts may be prevented, attenuated or compensated. Historic assessments provide a perspective on time as to the evolution of indicators, identifying trends which contextualize the present situation and the future scenarios.

The sustainability of electricity sectors has been carried out in other countries including Mexico [13], United Kingdom [14], Australia [15], Mauritius [16] and Turkey [17]. These investigations vary in terms of the scope considered when evaluating sustainability (only environmental or including economic and social components) and also in terms of the time extension (past, present and future projections).

The main objective of this paper was to evaluate the sustainability of Spain's electricity system. This assessment includes an investigation of historic data (since 1990) and future projections (2030 and 2050), which set a framework in which the current situation may be more adequately appraised. This transformation is evaluated on the basis of a series of indicators that describe the environmental, economic and socio-economic dimensions of the sustainability.

2. Methodology

2.1. Life Cycle Assessment Methodology

The sustainability of the Spanish electricity system has been calculated using three methodologies based on a life cycle approach: attributional life cycle assessment (LCA) for the environmental dimension, levelized cost of electricity (LCOE) for the economic dimension and direct life cycle employment generation for the socio-economic dimension. Figure 1 illustrates the scope and system boundaries applied in each one of these methodologies, which varied primarily depending on the availability of inventory data. Thus, the boundaries considered for the environmental dimension included the whole life cycle of the system, as included in the ecoinvent datasets employed [18]. The boundaries for the economic dimension considered all direct costs except decommissioning and the end of life phase [19,20]. The boundaries for direct employment generation only considered fuel generation (where necessary), manufacture and power plant construction and operation [21].

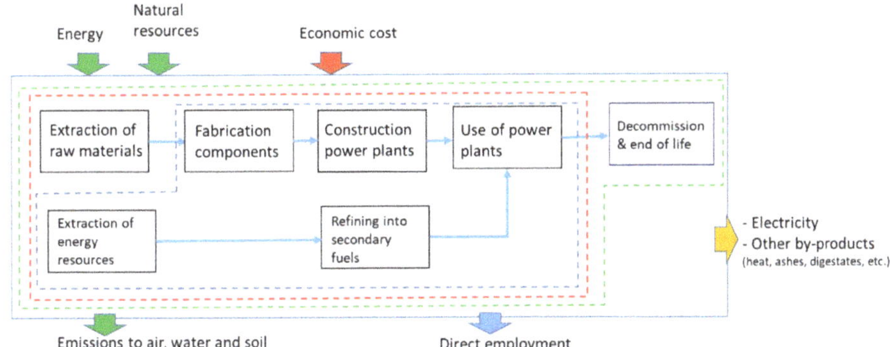

Figure 1. Life cycle diagram, system boundaries and input/output consideration employed to evaluate the environmental (green), economic (red) and socio-economic (blue) sustainability of Spain's electricity system.

2.2. Inventory Data of Installed Capacity, Electricity Generation and Technology Mix

The sustainability assessment of the electricity system was based on the quantification of indicators describing its life cycle performance on the environmental, economic and socio-economic dimensions. This required the collection of official information regarding installed capacity, power generation and technology mix for the time periods considered in the investigation (1990–2015 for historic analysis, and 2030 and 2050 for future projections).

2.2.1. Historic Electricity Data

The core of historic data for electricity generation, installed capacity and technology mix, covering the period 1990–2016, was extracted from La Energía en España, the official yearly report published by the Spanish Ministry for Energy, Tourism and Digital Agenda [22]. This information was validated and supplemented with additional data for 1960, 1970, 1980 and 1990 using updated statistics from Red Eléctrica Española (REE, the Spanish electricity system operator) and the International Energy Agency (IEA) [23,24]. Owing to the higher uncertainty associated with older generation technologies and their environmental, economic and socio-economic impacts, only the period 2000–2016 was evaluated for sustainability. Figure 2 illustrates this historic transformation in the Spanish electricity system in terms of electricity generation (GWh), generation per capita (MWh/capita) and technology mix for the period 1960–2016, and Figure 3 shows the same information for installed capacity for the period 2000–2016 (this parameter was not available for earlier years). The power generation technologies considered in this investigation are those listed in REE [23] and IEA [24] statistics as follows: Concentration Solar Power (CSP), Photovoltaic (PV), Wind, Hydropower (Hydro), Biomass, Nuclear, Oil, Natural gas and Coal. When different technological varieties are available for a given energy resource (e.g., natural gas in the form of combined cycles or CHP gas turbine, or wind in the form of off-shore and on-shore), the analysis is based on a weighted representation of the Spanish situation during the time period considered.

Generation values refer to gross power output, including electricity losses due to power transmission, distribution and other system inefficiencies. Due to its limited contribution, this investigation does not consider electricity imports and exports from Spain, which for the periods considered, accounted for between 2–3% of the power consumed nationally [22]. This investigation covers only power generation systems, overlooking other elements (e.g., storage, transformation, transmission, grid control) that may be essential in future electricity systems, particularly those with a strong dependence on renewables.

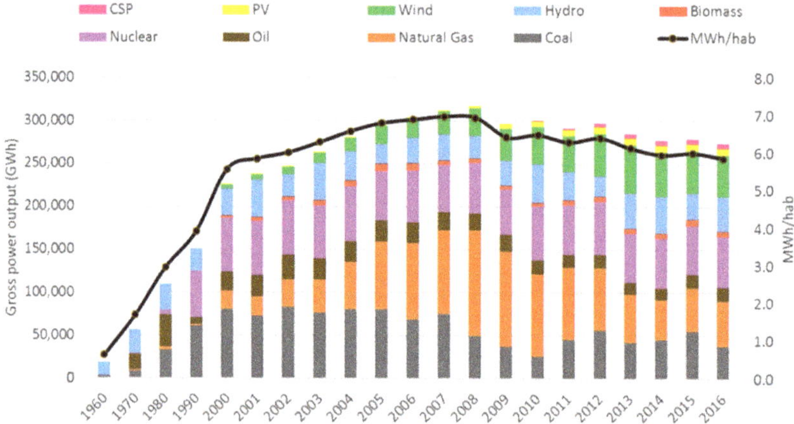

Figure 2. Historic data for electricity generation and technology mix in Spain (1960–2016).

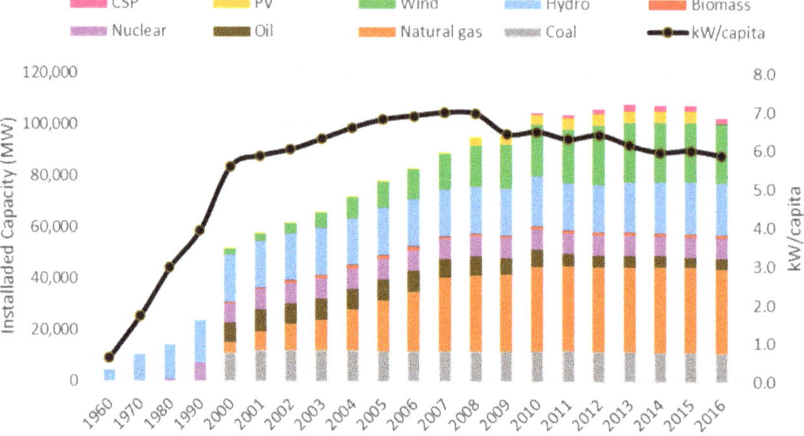

Figure 3. Historic data for installed generation capacity in Spain (2000–2016).

As shown in Figure 2, the commercial electrification of Spain commenced in the 1950s and 1960s with the development of the first large-scale hydroelectric projects. This was followed by a rapid expansion in the electricity sector between 1960 (18,615 GWh) and 2000 (300,777 GW/h), which was supported by the incorporation of oil, coal and also nuclear power to the technology mix. This electrification period was driven by the economic growth that followed the political transition into democracy in 1975 and the opening of the national markets that was culminated with the incorporation of Spain into the European Union in 1986. During this period, Spain developed most of its hydroelectric and nuclear capacity, which has remained rather stable up until the present (19.5 GW and 7.8 GW respectively by 2016).

After the year 2000, two different phases may be discerned. The first stage, between 2000 and 2008, is characterized by a progressive growth in power generation (35% increase from 225,000 GWh to 305,000 GWh) and, more notably, in installed capacity (91% increase from 51,000 MW to 97,500 MW) which aimed to provide stability to the national network. During this period, the technology mix was reinforced with a strong contribution of natural gas (both in terms of capacity and generation) and an incipient incorporation of renewables. The second stage, between 2008 and 2016, describes a less expansive and modernized economy where the power demand was rather stable or slightly decreasing due to the increase of sharing tertiary sector activities and the offshoring of energy intensive industrial activities. In terms of technology mix, that period sees a progressive expansion in the installed capacity and generation of renewables, primarily wind power (35% increase from 31,800 GWh to 48,900 GWh) and to a lesser extent PV (68.6% increase from 2500 GWh to 8000 GWh), CSP and biomass. Nuclear, oil and hydropower generation remained stable during that period, while coal and natural gas fluctuated to adapt to national strategies aimed at the promotion of national fuels (coal) and international commitments expected to tackle global warming [25,26].

2.2.2. Future Projections

The electricity projections investigated in this paper were defined by the think tank Economics for Energy and published in a document titled Scenarios for the Energy Sector in Spain 2030–2050 [27]. This data was revised and validated in a subsequent document titled Analysis and Proposals for Decarbonisation, commissioned by the Spanish Government and produced by the Commission of Experts on Energy Transition [28] The scenarios proposed incorporated the national objectives set under the Spanish Renewable Energy Action Plan 2020 [29], the international commitments assimilated in the ensuing European 2030 Climate and Energy Framework and the European 2050 long-term strategy [9,10].

Table 1 describes the scenarios analysed for sustainability, which are cited throughout the paper as follows: decarbonisation (DC), current policies (CP), accelerated technical advance (AT) and stagnation (ST). Figures 4 and 5 provides a graphical account in terms of power generation, installed capacity and technology mix.

Table 1. Summary of projected electricity scenarios for Spain, as extracted from [27,28].

Scenario	Year	Gross Generation (GWh)	Installed Capacity (MW)	Objectives
Current situation	2015	279,600	107,769	-
Decarbonization (DC)	2030 2050	290,653 477,073	141,968 247,324	Ambitious reduction of GHG emissions
Current Policies (CP)	2030 2050	310,997 416,698	112,757 219,979	Linear evolution of international geopolitics
Accelerated technology advance (AT)	2030 2050	352,260 581,930	153,787 290,764	Fast penetration of RE
Stagnation (ST)	2030 2050	281,460 352,507	108,399 108,755	High dependency on fossil fuels.

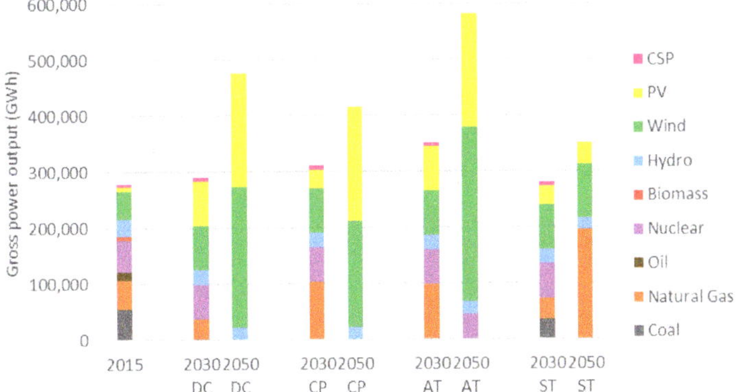

Figure 4. Projections of electricity generation and technology mix in Spain (2030–2050) compared to reference year 2015.

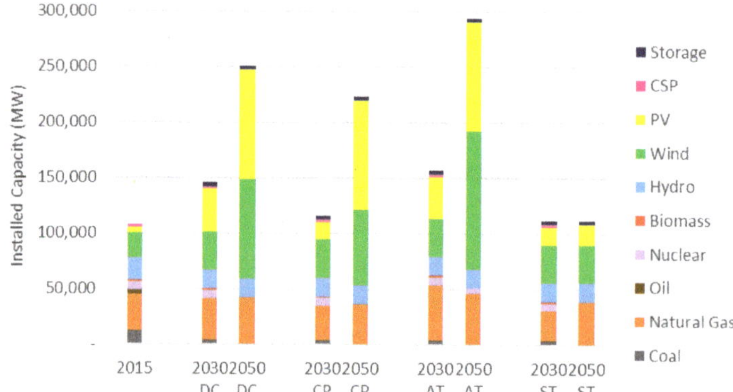

Figure 5. Projections of installed capacity and technology mix in Spain (2030–2050) compared to reference year 2015.

The decarbonisation (DC) scenario assumes the implementation of ambitious strategies to confront climate change and achieve a 40% reduction in GHG by 2030 and a 95% one by 2050, according to the objectives set by Member States of the European Union [9,10]. As shown in Figure 4, the DC scenario assumes a small increase in overall power consumption by 2030, as well as a complete elimination of coal and oil, a continuation of nuclear power and a slight reduction of natural gas from the electricity mix. This scenario also considers a significant increase in power demand by 2050, which is covered entirely by renewable sources, primarily PV and wind (hydropower remains stable due to limitations in the availability of additional hydroelectric resources in Spain).

The current policies (CP) scenario assumes a linear evolution of international geopolitics concerning the use of renewables and restraints in the emission of GHG. This scenario considers complete elimination of coal and oil from the electricity mix by 2030, assuming that this is largely replaced by natural gas, which absorbs 33% of the demand. By 2050, the CP scenario describes complete coverage of power demand from renewables, for an overall generation that is 15% lower than that in the DC scenario. The CP scenario would not accomplish the 95% GHG cuts proposed by the European Commission by 2050 [10].

The accelerated technology advance (AT) scenario presumes a rapid reduction in the cost of technologies related to renewable energies, energy storage and electricity consumption, including a fast transition into the electrification of transport. As shown in Figure 2, this would result in higher power demands than observed in the other scenarios and higher penetration of renewables as well as an achievement of the targets set by the European Commission for emission of greenhouse gases. The projections for this scenario consider elimination of coal and oil by 2030, which is compensated by a sustenance of nuclear power and a notable growth in natural gas and renewables that represent 54.5% of the mix. The predicted high power demands require a large penetration of PV and wind, and the continuation of nuclear energy in 2050.

The stagnation (ST) scenario considers a limited economic growth throughout this period and a limited development of new energy technologies leading to a time extension in fossil fuel dependence. In this scenario, overall demand remains fairly stable up until 2030, with a limited penetration of renewables and a strong presence of fossil fuels. The ST scenario assumes a prevalence of natural gas in the electricity mix by 2050 and a limited expansion of renewables in the long term.

2.3. Sustainability Factors of Power Generation Technologies

The sustainability assessment carried out in this investigation relies on a series of emission, economic and employment factors defined for the life cycle of each of the generation technologies that compose the national electricity mix. This section describes the methodologies followed to define these factors.

2.3.1. Environmental Dimension

Environmental emissions associated with individual power generation technologies were extracted from Ecoinvent v3.1 [30]. The inventory data in these datasets cover the following life cycle stages: (i) extraction and processing of raw materials employed in the construction of power generation infrastructures, (ii) construction of power plants, end of life of construction materials, extraction and processing of fuels (where required); and (iii) operation of power plants and power transmission. When more than one dataset was available for any given technology, a weighted combination of the situation describing the Spanish electricity system was employed. Since no background data was available for CSP plants, and due to the fact that its contribution to the Spanish electricity system is limited (4.0% in 2015), the emissions associated with this technology were not considered.

Environmental impact assessment calculations were carried out using the ILCD 2011 Midpoint+ method [31], except for the human toxicity category for which the ReCiPe 2016 Midpoint (H) v1.03 method [18] was used. This latter method was favoured over ILCD in the human toxicity category as it provided an aggregated approach that included both cancerous and non-cancerous effects.

The environmental categories and the impact units considered in this investigation include the global impacts: climate change (kg CO_2 eq), fossil depletion (kg oil eq) and ozone layer depletion (g CFC-11 eq), and the more locally focused human toxicity (kg 1.4 DB eq), terrestrial acidification (kg SO_2 eq) and photochemical ozone formation (kg NMVOC eq).

2.3.2. Economic Dimension

The levelized cost of electricity (LCOE) was employed to evaluate the economic sustainability of the Spanish electricity. This indicator has a life cycle approach that is calculated by dividing the discounted cost of power generation (including investment, operation and maintenance, fuel expenditures and decommissioning) by the discounted rate of power generation, as shown in Equation (1):

$$LCOE\left(\frac{€}{MWh}\right) = \frac{discounted\ lifetime\ costs}{discounted\ power\ generation} = \frac{\sum_{t=1}^{n} \frac{I_t + M_t + F_t}{(1+r)^t}}{\sum_{t=1}^{n} \frac{E_t}{(1+r)^t}} \tag{1}$$

where I_t, M_t and F_t represent investment, operations and maintenance and fuel expenditures in the year t, and E_t represents the power generated in the same year t. The value r represents the discount rate assumed for the power generation project and n its expected lifetime.

The LCOE considered for each of the technologies considered in the Spanish electricity mix were obtained from the International Energy Agency [20] for scenarios prior to 2016. LCOE values were calculated assuming a discount rate of 7.0% and had a national specificity. In cases where this information was not available for Spain (e.g., coal, natural gas combined cycle and nuclear), the cost values were calculated as the average of those applicable to countries within the European Union. Additional information about other key parameters (e.g., technology type and lifetime, average capacity factors) employed to calculate the LCOE may be found in [20].

The future cost of power generation technologies is a matter of debate [32–37]. For the purpose of this investigation, a dynamic approach has been applied based on a series of factors applicable to the reference costs proposed by the International Energy Agency [20]. The transformation factors used for the period 2015–2030 were those proposed by [33] as follows: coal (−5.43%); natural gas (+46.02%); nuclear (+9.51%); hydro (−27.49%); wind (−54.30%); PV (−55.28%) and CSP (−56.95%). Reliable transformation factors for the period 2015–2050 were only available for wind (−69.93%) [34] and PV (−64.22%) [35], which are the most dominant technologies in all the 2050 scenarios (except for ST, which incorporates a high proportion of natural gas). In the absence of dependable data for other technologies, the costs assumed for nuclear, natural gas and hydro in 2050 were the same as in 2030. In view of past trends (increasing costs of fossil resources and reduced costs for renewables), these assumptions are likely to represent an underestimation in the cost of natural gas and nuclear power.

Figure 6 shows the LCOE applied to different power generation technologies in Spain, according to the procedures described in [20]. The figures show that CSP, waste incineration and biomass have the highest costs. The cost of renewables (wind, PV and hydro) is comparable to conventional fossil fuels and nuclear energy, with coal power being the cheapest. The cost of fossil technologies is dominated by the operation phase, due to the expenses associated with the extraction and processing of fuels, while the cost of renewables is dominated by the construction of the infrastructures allocated to the capital costs. Certain technologies (biomass, CHP, biogas) benefit from heat credits due to the combined generation of power and thermal energy. To avoid the results being affected by international currency policies, a fixed exchange rate was used to convert monetary data published by IEA from USD to Euro. The exchange rate considered was the average value for the core assessment dates (2010–2015) as reported by the European Central Bank at 1 USD = 0.77 €.

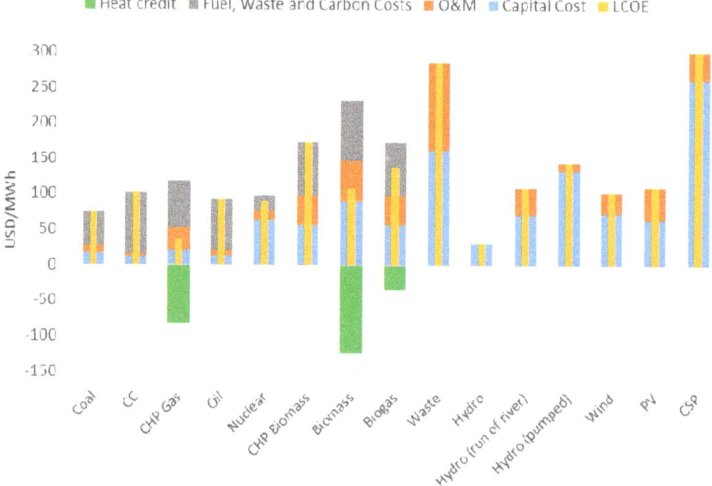

Figure 6. Economic performance of different technologies for power generation, as applicable to Spain in 2015 in terms of LCOE (adapted from [20]).

This economic analysis does not take into consideration external costs in the form of carbon taxes or carbon emission credits. The incorporation of these levies would be particularly detrimental to the economic interest of the scenarios with a higher contribution of fossil technologies.

2.3.3. Socio-Economic Dimension

The socio-economic performance of the power system in Spain was evaluated using the direct employment generated by the technologies participating in the electricity mix as the indicator. The methodology employed was published by the Institute for Sustainable Futures at the University of Technology Sydney (ISF-UTS) and follows a life cycle approach that takes into consideration four stages: extraction of raw materials and manufacturing of components; construction and installation of additional capacity; operation and maintenance of power plants; and extraction and refining of fuels (where necessary).

Data from the original report published in 2010 [38] was used to quantify employment in the Spanish electricity sector prior to 2010, data published in a subsequent update from 2012 [39] was used to quantify the period 2012–2013 and data from the latest report of 2016 was used to quantify the period 2014–2016. As technology becomes more mature, employment requirements decrease. The employment reduction factors proposed by [21] were used to quantify the situation in the projected scenarios of 2030 and 2050. This methodology takes into consideration the geographic location of the energy projects and time factors that account for expected deviations in future scenarios (due to learning curves and economy of scale). Figure 7 shows the employment factors used for the calculation of jobs in the Spanish electricity systems, as extracted from the references cited above.

Figure 7. Employment factors used to quantify direct jobs created from different power generation technologies (adapted from [21,38,39]).

The construction periods selected for different technologies in these calculations were as follows: 10 years for nuclear power plants, five years for coal, two years for natural gas, oil, biomass, hydroelectric, wind, CSP and combined heat and power (CHP), and one year for PV, as reported by [21].

3. Results and Discussion

This section describes the evolution in the sustainability indicators as calculated for the Spanish electricity system since 1990 and also for the alternative scenarios projected for 2030 and 2050. The year 2015 has been used as a reference for the current situation. As described above, the sustainability indicators cover three dimensions (environment, economic and socio-economic), which are structured into the following three sub-sections.

3.1. Environmental Sustainability Assessment

Figure 8 illustrates the evolution in the environmental sustainability of the Spanish electricity system in six categories. The analysis of the global warming category (top left) shows a rapid increase in total GHG emissions since 1990 (7.9×10^{10} t CO_2/year) to reach a maximum in 2007 of 1.72×10^{11} t CO_2/year. This is caused primarily by the progressive growth in the economic activity of the country which demands increasing consumptions of electricity for industrial and domestic applications, as illustrated in Figure 2. During this period, the carbon intensity of the electricity system (GHG emissions per unit of power generated) remains rather stable at values between 0.50–0.57 t CO_2/MWh, due to the strong contribution of fossil fuels, primarily coal and oil.

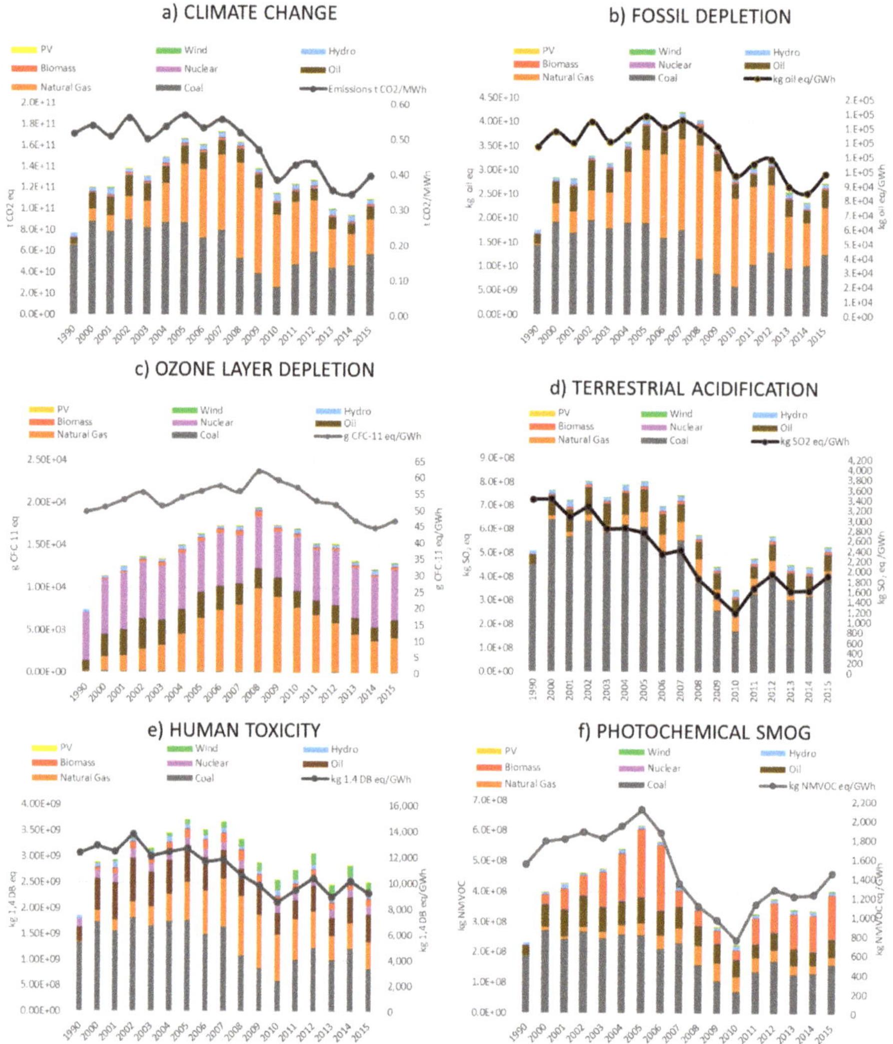

Figure 8. Historic environmental performance of Spain's electricity system (1990–2015): (**a**) climate change, (**b**) fossil depletion, (**c**) ozone layer depletion, (**d**) terrestrial acidification, (**e**) human toxicity, (**f**) photochemical smog.In the first stage of the ST scenario (2030), the results show a notable reduction (44%) in the emissions per unit of power, due to a decline in the contribution of coal, the complete elimination of oil and the increasing weight of renewables (mainly PV). In the second stage of the ST scenario (2050), the higher impact per unit of power compared with 2030 is associated to the prevalence of natural gas in the power mix and the not so decisive upsurge in the penetration of renewables.

The carbon footprint of the power system is progressively reduced after 2008 reaching a minimum of 9.6 t CO_2/year in 2014. This situation may be associated with the economic recession that the Spanish economy suffered between 2011 and 2013 due to the global financial crisis. By looking at the carbon intensity, the results evidence a progressive reduction in the emissions per unit of power between

2007 (0.56 t CO_2/MWh) and 2014 (0.38 t CO_2/MWh). This is attributable to the implementation of the national strategy endorsing the use of natural gas (burnt in higher efficiency and lower carbon intensity combined cycles) and the onset of ambitious national plans for the promotion of renewables. These two approaches progressively relegated the use of the more carbon intensive coal and oil power plants. A detailed description of the aggressive policies implemented by the Spanish Government in that period to favour the deployment of renewables may be revised in [40,41]. The progressive recovery of the Spanish economy and a renewed interest in the use of national coal for power generation explains the gradual upturn in GHG emissions after 2014, despite the increasing contribution of renewables.

A similar approach may be used to evaluate the evolution in the environmental performance of the Spanish electricity system on the other five categories considered in this investigation. In all cases, this performance may be related to changes in the overall demand and technology mix of the system. As explained in Figures 2 and 3, this is conditioned by the economic and political situation of the county, with a rapid economic growth and electrification based on fossil fuels between 1990 and 2000, and a more progressive increase in demand (typical of a more mature economic situation) based primarily on natural gas from 2008. From this date until the present, there is a gradual reduction in power generation partially attributable to the global economic crisis and also to the reinforcement of the tertiary sector and delocalization of more energy intensive economic activities. This period is also characterized by a progressive but strong public support for renewables. In order to avoid extending this section excessively, the discussion on the remaining environmental categories is more restrained, focusing primarily on general trends and key issues.

Thus, in the category describing the depletion of fossil resources (top middle), the results show a profile very similar to that of fossil fuel utilization. The small contribution of hydropower to this category is related to fossil fuel utilization and the consumption of other natural resources during the construction phase of the plants (mainly reservoirs). Regarding the ozone layer depletion category, the emissions follow a pattern strongly affected by the use of nuclear and natural gas power. This is due to the emission of halogenated hydrocarbons (Freon, Halon, CFCs, HCFCs, etc.) used as refrigerating and fire suppressing agents.

The results show a solid correlation between terrestrial acidification (bottom left) and fossil fuel utilization, primarily coal. Thus, the progressive reduction in coal contribution between 1990 and 2010 results in lower impact values in this category. Changes in the Spanish policies regarding the promotion of national coal or its substitution for natural gas and renewables, due to European commitments related to climate change, are responsible for the fluctuations in the acidification impact observed after 2010.

In terms of human toxicity, the assessment of the Spanish electricity system illustrates a progressive reduction in the emissions of 1.4 DB eq per unit of power between 1990 and 2010, later followed by a certain degree of stabilization. The first period may be explained by a progressive penetration of natural gas (at the expense of coal and oil) and the second by the significant contribution of renewables to this category. Regarding the photochemical smog category, the results evidence the strong contribution of coal and biomass plants, or rather, the comparatively smaller contribution to this category of all other power generation technologies. Therefore, the results show an increase in NMVOC emissions per unit of power generated between 1990 and 2005 due to the incorporation of biomass power plants. This is followed by a rapid reduction between 2006 and 2011 due to the smaller contribution of biomass and coal to the power mix, followed by an upturn after 2012 due to an increase in power demand and the incorporation of additional biomass capacity.

Figure 9 describes the same environmental profiles generated by the Spanish electricity system as determined in the four scenarios of 2030 and 2050. Regarding the climate change category, the results evidence a reduction in GHG emission in all cases, which is less marked in the stagnation (ST) scenario due to the prevalence of fossil fuels in the electricity mix.

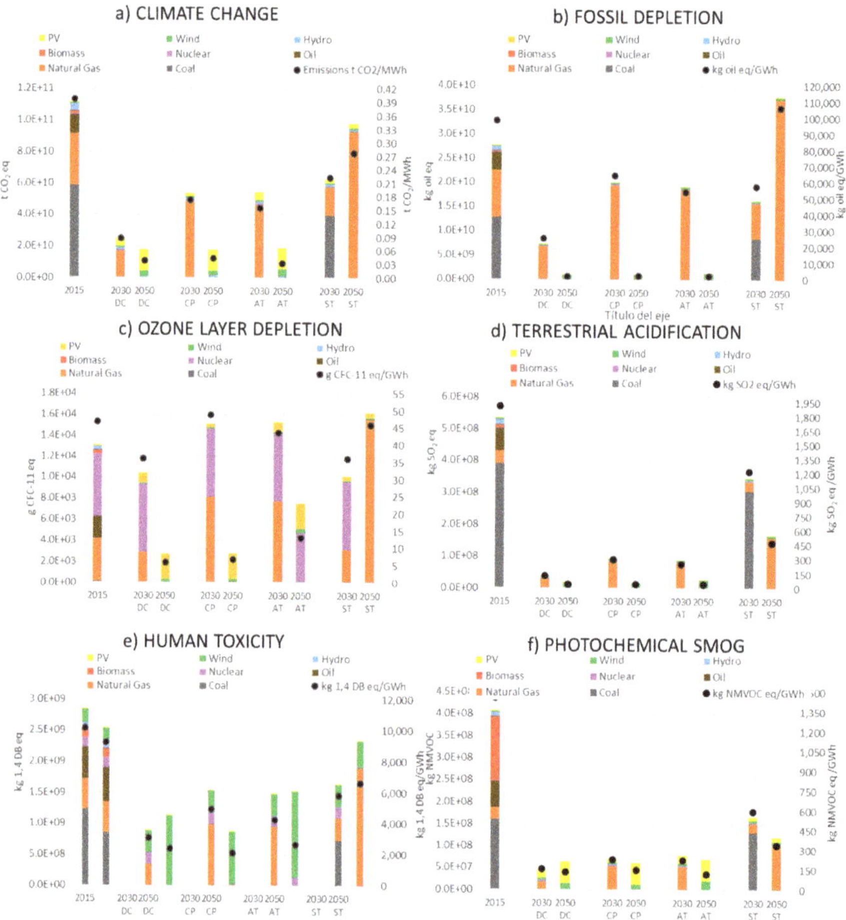

Figure 9. Environmental performance of future projections of Spain's electricity system: 2030–2050: (**a**) climate change, (**b**) fossil depletion, (**c**) ozone layer depletion, (**d**) terrestrial acidification, (**e**) human toxicity, (**f**) photochemical smog.

On the opposite side of the spectrum, the decarbonisation scenario (DC) shows a very significant reduction (78%) in GHG emissions by 2030 due to complete elimination of coal and oil from the mix, and the strong contribution of renewables. This trend continues up until 2050 where all the power is generated by renewables (mainly PV and wind, with a small contribution from hydropower), resulting in a very reduced overall GHG output, both in total terms (2.4×10^{10} t CO_2 eq) and per unit of power generated (0.037 t CO_2 eq/MWh).

The other two scenarios (CP and AT) show similar patterns to each other in terms of overall GHG emissions. The impact on climate change generated in 2030 is expected to be less severe than that calculated in the DC scenario, due to the comparatively higher contribution of natural gas, which is used to smooth the transition towards the elimination of fossil fuels. In the longer term (2050), the three scenarios (DC, CP, AT) generate a similar impact in the climate change category. However, since the AT assumes a higher power demand, this results in a smaller impact per unit of power (0.033 t CO_2 eq/MWh), which is achieved by permitting a certain contribution of nuclear up until 2050.

A simplified analysis of the situation regarding the other five impact categories and the four projected scenarios is provided below. Regarding the fossil depletion category, the results show a reduction in each scenario, except for ST, due to the substitution of fossil fuels for renewables. The benefits are more markedly observed in 2030 in the DC scenario, due to the more ambitious stance on renewables, compared to CP and AT. In 2050, the impact generated on this category was insignificant in the three scenarios that opted for renewables (DC, CP and AT), but even higher than the present situation in the ST scenario, due to its strong dependence on natural gas.

With regards to ozone layer depletion, the results show a strong dependence on the use of nuclear power and natural gas. These technologies are favoured in all four future scenarios in 2030, which is why impact values on these categories are not reduced in this time horizon. However, in the longer term (2050), the results evidence a significant reduction in the scenarios that assume the closure of nuclear power and the elimination of natural gas (DC and CP). In contrast, the stagnation (ST) scenario maintains a very high impact due to its reliance on natural gas.

Regarding terrestrial acidification, the results show a strong alleviation in this impact category in each scenario, except ST, due to elimination of coal from the mix. This effect is more marked in the DC scenario due to the more decisive penetration of renewables and elimination of sulphur containing fuels (mainly coal but also oil and natural gas). In terms of human toxicity, the results show a notable reduction both in gross emissions and per unit of power generated. These benefits are less marked in the ST scenarios due to the strong contribution of coal in 2030 and natural gas in 2050. The reduced impact in this category observed in 2030 in the DC scenario (compared against CP and AT) is due to the limited contribution of natural gas. The total impact on this category is still noticeable in 2050 due to toxic emissions associated with the life cycle of renewables, primarily in their fabrication stage.

Finally, the impact generated in the photochemical smog category is significantly reduced in each scenario due to the elimination of biomass and coal power plants (except ST in 2030). The use of natural gas to smooth the transition into renewables in the CP and AT scenarios is responsible for the higher impact on this category in 2030 and that is also why ST scenario shows a higher impact in this category of the ST scenario, compared to the other three. Of the other three, AT showed the lowest impact per unit of power in 2050 due to the higher power demand assumed and the utilization of nuclear stations.

The economic sustainability of the Spanish electricity system has been evaluated using the LCOE as the indicator. This analysis only covers the period 2010–2016 due to lack of information regarding the installation of additional capacity in earlier years. Figure 10 shows that the total cost of power generation in Spain in 2010 is dominated by fossil and nuclear technologies. The generation of renewables grows progressively in this period and so does their economic contribution to the electricity system. Overall, the total cost of electricity in Spain during this period (2010–2016) does not change significantly, due to the fact that the renewables with a higher contribution (wind and PV) are attributed similar costs to fossil technologies, while the contribution of higher cost renewables (CSP, biomass and biogas) is still marginal (see Figure 6).

3.1.1. Electricity Costs in Historic and Present Data

Figure 11 shows a breakdown of the total cost of power generation in Spain during the reference year 2015, indicating the contribution of different technologies and life cycle stages.

The results show that most of the LCOE in the Spanish electricity system correspond to the capital cost of building the infrastructures (52%), followed by fuel costs (37%) and to a lesser extent the operation and management of the plants (11.1%). The main contributors to the capital costs are the nuclear energy and the renewable technologies (primarily wind). It is also remarkable the relatively high contribution of CSP to overall power costs, despite its limited generation share. Regarding fuel costs for 2015, the results show a large contribution of natural gas and a smaller proportion of coal, oil and nuclear. The contribution of biomass costs is very limited but still far greater than what should correspond to its limited generation capacity. The contribution of other renewables to this life cycle stage is obviously null.

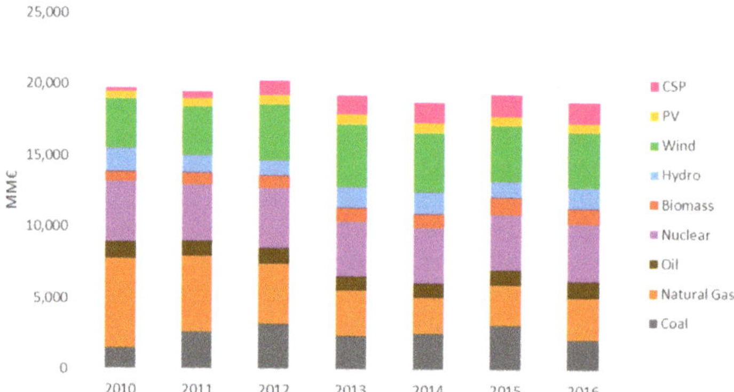

Figure 10. Historic values of power generation costs estimated as a summation of LCOE of all contributing technologies.

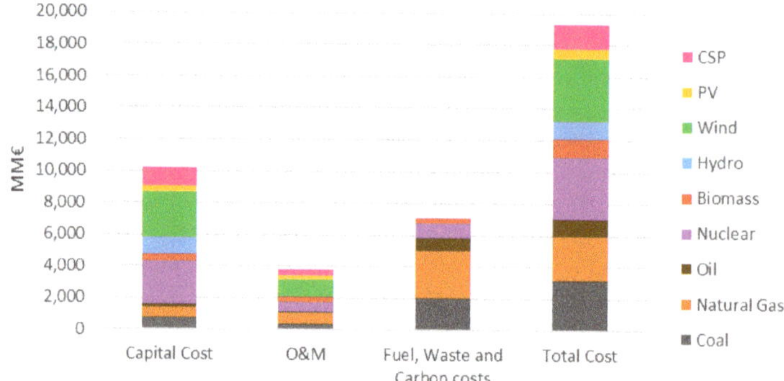

Figure 11. Contribution of different technologies to the aggregated LCOE of Spain in 2015

3.1.2. Electricity Costs in Future Projections

Figure 12 illustrates the cost of power generated in the four scenarios projected for 2030 and 2050. The results show a cost reduction per unit of energy generated (€/MWh) in each projected scenario when compared to the reference year of 2015 (69.12 €/MWh). Cost cuts grow larger between 2030 and 2050 in each scenario. These cuts are greater in the scenarios dominated by renewables (DC = 29.76 €/MWh; CP = 30.63 €/MWh; AT = 39.77 €/MWh in 2050) due to the cost reductions envisaged for wind and PV. The cost differences between the scenarios dominated by renewables and fossil technologies are less marked in the short term (2030) but become remarkable in the long term (2050) scenarios. Despite being slightly lower than that of 2015, the cost per unit of power of the scenarios with the strongest contribution of renewables (ST) is by far the highest of all in the long run (57.74 €/MWh in 2050).

Comparing the overall cost of the electricity systems is less apparent due to differences in the power demand considered in each scenario (see Figure 4). The results evidence a progressive cost reduction in the decarbonization (DC) and current policy (CP) scenarios when compared to the situation in 2015. This is so despite the significantly higher generation values considered in the future scenarios (DC = 477,073 GWh and CP = 416,698 GWh in 2050, compared to the 279,600 GWh for 2015). This is due to the strong penetration of renewables and the cost reductions envisaged for wind and solar. The higher overall cost generated by the AT scenario is due to the strong power demand associated with this case (581,930 GWh in 2050, almost double of that in 2015). Despite assuming the lowest generation values (ST = 352,507 GWh in 2050) the overall cost of the stagnation scenario was one of the highest due to the high economic intensity of fossil fuels and the limited contribution of renewables.

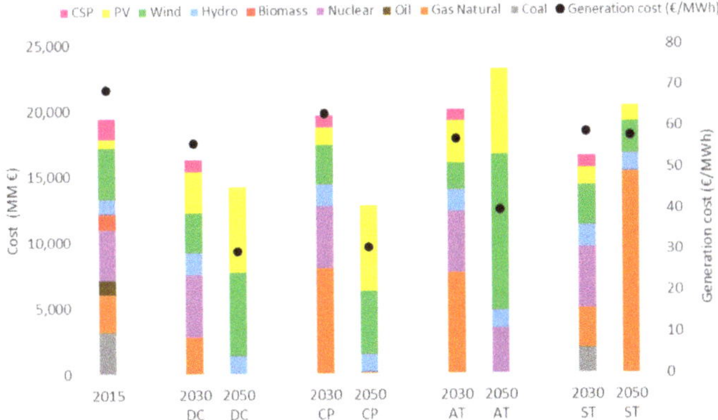

Figure 12. Power generation costs (LCOE) and contribution of different technologies in the four scenarios considered for 2030 and 2050 (DC = decarbonization; AP = maintaining current policies; AT = Advanced technologies; ST = stagnation).

3.2. Socio-Economic Sustainability Assessment

This section revises the evolution of direct employment generation in the Spanish electricity system over the last decades and the projections for the future scenarios of 2030 and 2050. Figure 13 shows employment generation in the Spanish electricity system between 2010 and 2016. As explained for the economic assessment, the employment evaluation only covers the period 2010–2016 due to lack of information regarding the installation of additional capacity in earlier years.

Figure 13. Historic data (2010–2016) describing employment generation of Spain's electricity system per technology.

3.2.1. Employment Generation in Historic and Present Situation

The results in Figures 13 and 14 show high employment generation rates (86,000 jobs) in 2010, with a very strong contribution of renewables, primarily in the manufacturing and construction stage. This reflects the rapid expansion in the installed capacity of renewable energies (primarily wind and PV, but also CSP) that occurred in Spain during the first years of the 2010 decade [42,43]. Most of the jobs generated by the PV sector are associated with the installation (construction) of the plants while most of the jobs generated by wind are related to the manufacturing of components.

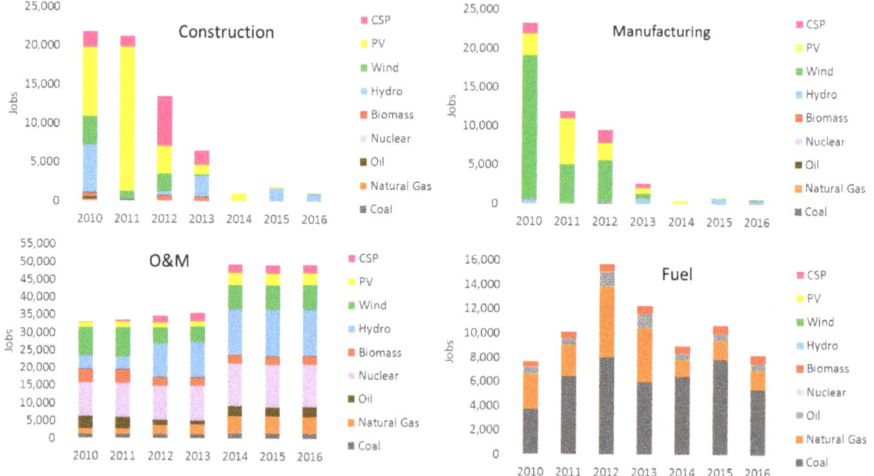

Figure 14. Employment generation in the Spanish electricity system since 2010 and disaggregated per life cycle phases: construction and installation of power plants (top left), raw materials and manufacturing of components (top right), operation and management of power plants (bottom left) and fuel production (bottom right).

The results show a progressive reduction in the generation of jobs between 2010 and 2013 in the manufacturing and construction phases, due to the progressive stagnation of the renewable sector and the stoppage in the deployment of additionally installed capacity. The exception to this trend is CSP, which peaks in employment generation during 2012 and 2013 due to the installation of new plants. The inactivity in the construction of new renewable plants extends until 2016, incorporating CSP after 2013.

Regarding the operation and management, the costs remain rather stable between 2010 and 2013. Then there is a jump between 2013 and 2014 and then they are stable again until 2016. In this life cycle stage, most jobs go to nuclear and hydroelectric (around 10,000 jobs each). It should be noted that jobs in operation and management are more stable than those in the construction stage, which are temporary as they last as long as the construction of the plant takes.

In contrast, the number of jobs attributable to the extraction and processing of fuels follows the trail marked by fossil technologies, with a relative maximum in 2012 which is the year with the highest contribution of coal power. Biomass is also a highly employment intensive technology, although its contribution to the Spanish electricity mix is very limited. Despite the lower contribution to power generation, most of the jobs in this category are attributable to coal, followed by natural gas, depending on its contribution to the mix.

Figure 15 illustrates the total number of direct jobs generated by the electricity system in Spain in the reference year of 2015. The power costs in this time are strongly dominated by the operation and management phase (78.43%), due to the very limited additional capacity projected for this year and the already strong contribution of nuclear power and renewables.

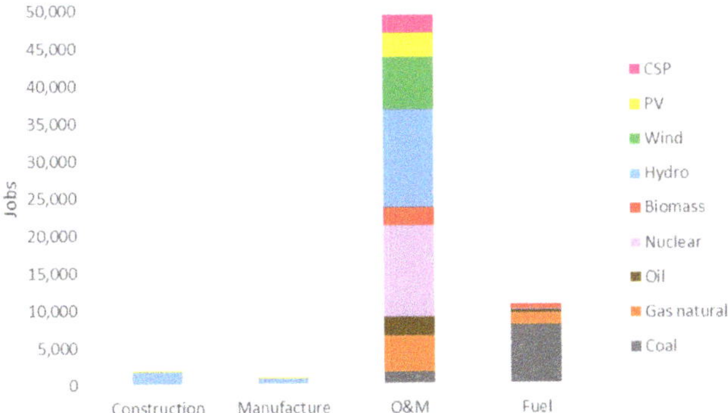

Figure 15. Breakdown of direct employment generation by power generation technology (2015).

3.2.2. Employment in Future Projections

Figure 16 shows a comparative analysis of employment generation in the four future scenarios considered in this investigation. The results represent accumulated employment for each of the evaluated periods (2015–2030 and 2030–2050), including that associated with the installation of additional capacity (construction and manufacturing), operation and management, and fuel extraction and processing (where necessary). The results show higher employment rates in the three scenarios that assume a stronger penetration of renewable technologies (DC, CP and AT). In all these cases, employment is primarily associated with the deployment of PV and wind energy. This employment is generated earlier (2030) in the scenarios assuming a rapid transformation of the electricity model (DC and AT), and later (2050) in those assuming a more gradual conversion (CP). Overall employment generation is greatest in the scenario that proposes a higher overall power demand (AT). In contrast, the scenario that assumed a strong dependence on fossil fuels (ST) shows the lowest job gains, most of which are still related to the mild deployment of PV and wind power.

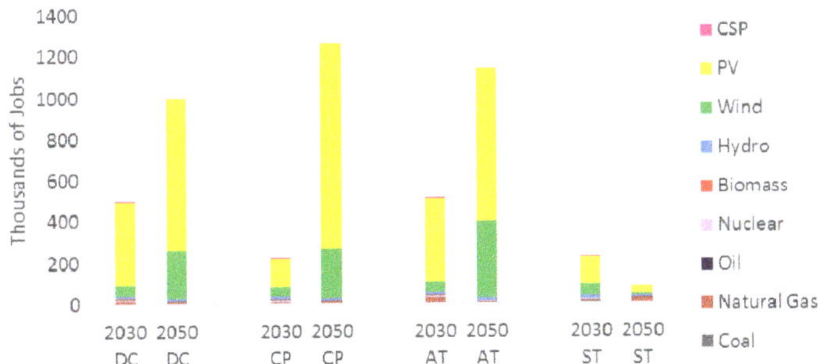

Figure 16. Employment generation of Spain's electricity system: future projections.

As an example, Figure 17 shows the distribution of jobs throughout the life cycle of the technologies contributing to the mix. This exercise has been done for the AT scenario in 2050, although the same discussion may be applicable to describe the situation in the other two scenarios describing a strong dependence on renewables (DC and CP). Thus, the results show most of the jobs that will be generated will be related to the manufacturing of the components and installations, and primarily to the construction of the new power plants. These jobs will be absorbed primarily by the PV technology and, to a lower extent, wind technology. The O&M phase has a significantly lower contribution to employment generation, while the contribution of fuel generation and transformation is negligible.

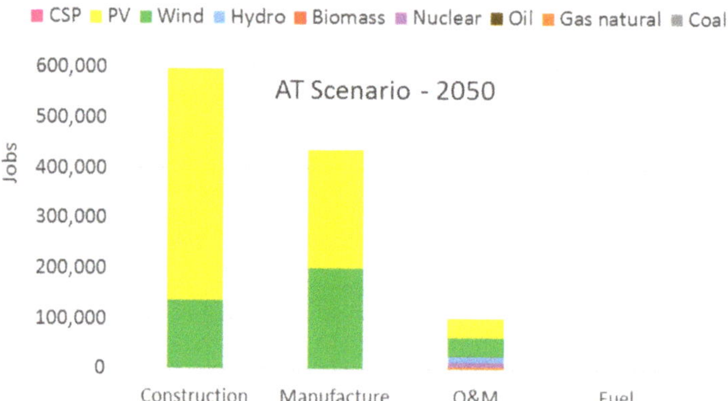

Figure 17. Breakdown of direct employment generation by life cycle phase (construction, manufacturing, Operation and Management, and fuel) in 2050 for the AT scenario.

4. Conclusions

The electricity system of a country has enormous repercussions on its sustainability. This investigation describes the evolution in environmental, economic and socio-economic sustainability of the Spanish electricity system between 1990 and 2015, and also in four future scenarios projected for the years 2030 and 2050.

The results have shown that between 1990 and 2000, there is a strong increase in the impacts generated by the system on most environmental categories. This is due to the fact that this period was characterized by solid economic growth which caused a robust demand for this energy vector, and also a strong dependence on fossil fuels of the technology mix. The total cost of power generation also escalates rapidly and so does the generation of employment.

In the period between 2000 and 2008, the results show a progressive but less rapid increase in power demand which is met using natural gas, while coal electricity reduces its predominance gradually. This results in reduced impact values in global categories like climate change, a limited effect on electricity generation cost and a stabilization in the generation of employment.

The period between 2008 and 2016 combines a strong economic crisis with an ambitious public strategy aimed at promoting renewables. The result is a progressive reduction in the impacts associated with global warming, a slight increase in the generation costs and a notable increase in employment generation.

Regarding the future projections, the results show that the scenarios with a higher contribution of renewables (DC, CP, AT) exhibited reduced GHG emissions per unit of power and achieved higher employment rates, all while having a lower economic cost. These benefits become more noticeable in the longer term (2050). The opposite is observed in the scenarios that assume a higher dependency on fossil technologies (ST).

Author Contributions: Conceptualization, G.S.M.; methodology, G.S.M. and M.C.; formal analysis, M.C.; investigation, G.S.M. and M.C.; data curation, G.S.M. and M.C.; writing—original draft preparation, G.S.M.; writing—review and editing, G.S.M.; visualization, G.S.M. and M.C.; supervision, G.S.M. All authors have read and agreed to the published version of the manuscript.

Funding: This piece of research received no external funding.

Conflicts of Interest: The authors declare no conflict of interest.

References

1.   World Bank. *State of Electricity Access Report 2017*; World Bank: Washington, DC, USA, 2017; ISBN 1202522262.

2.   United Nations. *Sustainable Development Report 2019: Transformations to Achieve the Sustainable Development Goals*; United Nations: New York, NY, USA, 2019.

3.   Nuclear Energy Agency. *The Full Costs of Electricity Provision*; OECD: Paris, France, 2018; ISBN 9789264303119.

4.   IRENA. *Renewable Power Generation Costs in 2017*; International Renewable Energy Agency: Abu Dhabi, UAE, 2018; ISBN 978-92-9260-040-2. Available online: http://irena.org/publications/2018/Jan/Renewable-power-generation-costs-in-2017 (accessed on 12 February 2020).

5.   Turconi, R.; Boldrin, A.; Astrup, T. Life cycle assessment (LCA) of electricity generation technologies: Overview, comparability and limitations. *Renew. Sustain. Energy Rev.* **2013**, *28*, 555–565. [CrossRef]

6.   Barros, M.V.; Salvador, R.; Piekarski, C.M.; de Francisco, A.C.; Freire, F.M.C.S. Life cycle assessment of electricity generation: A review of the characteristics of existing literature. *Int. J. Life Cycle Assess.* **2020**, *25*, 36–54. [CrossRef]

7.   WG Environmental Management & Economics. Eurelectric, Life Cycle Assessment of Electricity Generation. 2011. Available online: http://www.eurelectric.org/media/26740/report-lca-resap-final-2011-420-0001-01-e.pdf (accessed on 15 July 2019).

8.   European Commission. *Directive 2009/28/EC of The European Parliament and of the Council of 23 April 2009 on the Promotion of the Use of Energy from Renewable Sources*; European Commission: Brussels, Belgium, 2009.

9.   European Commission. EU Climate Action. Available online: https://ec.europa.eu/clima/policies/eu-climate-action_en (accessed on 15 February 2020).

10.   European Commission. *A Clean Planet for all a European Strategic Long-Term Vision for a Prosperous, Modern, Competitive and Climate Neutral Economy*; COM/2018/773 final 2018; European Commission: Brussels, Belgium, 2018.

11.   IPCC. *Climate Change 2014: Synthesis Report*; Contribution of Working Groups I, II and III to the Fifth Assessment Report of the Intergovernmental Panel on Climate Change; Pachauri, R.K., Meyer, L.A., Eds.; IPCC: Geneva, Switzerland, 2014.

12.   IRENA. *Rethinking Energy 2017*; IRENA: Abu Dhabi, UAE, 2017; Volume 55, ISBN 978-92-95111-06-6.

13.   Santoyo-Castelazo, E.; Stamford, L.; Azapagic, A. Environmental implications of decarbonising electricity supply in large economies: The case of Mexico. *Energy Convers. Manag.* **2014**, *85*, 272–291. [CrossRef]

14.   Azapagic, A.; Chalabi, Z.; Fletcher, T.; Grundy, C.; Jones, M.; Leonardi, G.; Osammor, O.; Sharifi, V.; Swithenbank, J.; Tiwary, A.; et al. An integrated approach to assessing the environmental and health impacts of pollution in the urban environment: Methodology and a case study. *Process Saf. Environ. Prot.* **2013**, *91*, 508–520. [CrossRef]

15.   May, J.R.; Brennan, D.J. Sustainability Assessment of Australian Electricity Generation. *Process Saf. Environ. Prot.* **2006**, *84*, 131–142. [CrossRef]

16.   Brizmohun, R.; Ramjeawon, T.; Azapagic, A. Life cycle assessment of electricity generation in Mauritius. *J. Clean. Prod.* **2015**, *106*, 565–575. [CrossRef]

17.   Atilgan, B.; Azapagic, A. An integrated life cycle sustainability assessment of electricity generation in Turkey. *Energy Policy* **2016**, *93*, 168–186. [CrossRef]

18.   Huijbregts, M.A.J.; Steinmann, Z.; Elshout, P.M.; Stam, G. ReCiPe2016: A harmonised life cycle impact assessment method at midpoint and endpoint level. *Int. J. Life Cycle Assess.* **2016**, *22*, 138–147. [CrossRef]

19.   IEA. *Investment Costs for Power Generation*; International Energy Agency: Paris, France, 2017; Available online: http://www.worldenergyoutlook.org/weomodel/investmentcosts/ (accessed on 17 June 2019).

20.   IEA. *Projected Cost of Generating Electricity, International Energy Agency (IEA), Organization for Economic Cooperation and Development—Nuclear Energy Agency (OECD-NEA)*, 2015 edition; IEA: Paris, France, 2015.

21.   Rutovitz, J.; Dominish, E.; Downes, J. *Calculating Global Energy Sector Jobs: 2015 Methodology*; Institute for Sustainable Futures, University of Technology Sydney: Sydney, Australia, 2015.

22.   MINETAD. La Energía en España. 2016. Available online: https://energia.gob.es/balances/Balances/LibrosEnergia/energia-espana-2016.pdf (accessed on 16 June 2019).

23.   REE. Series Estadísticas Nacionales. Available online: https://www.ree.es/es/datos/publicaciones/series-estadisticas-nacionales (accessed on 16 June 2019).

24.   IEA. Energy Statistics. 2018. Available online: https://www.iea.org/countries/spain (accessed on 16 June 2019).

25.   Zafrilla, J.E. The mining industry under the thumb of politicians: The environmental consequences of the Spanish Coal Decree. *J. Clean. Prod.* **2014**, *84*, 715–722. [CrossRef]

26. Bianco, V.; Driha, O.M.; Sevilla-Jiménez, M. Effects of renewables deployment in the Spanish electricity generation sector. *Util. Policy* **2019**, *56*, 72–81. [CrossRef]

27. Linares, P.; Declercq, D. Escenarios Para el Sector Energético en España, 2030–2050. 2018. Available online: https://eforenergy.org/docpublicaciones/informes/informe_2017.pdf (accessed on 16 June 2019).

28. Government of Spain. Comisión de Expertos de Transición Energética. 2018. Available online: http://www6.mityc.es/aplicaciones/transicionenergetica/informe_cexpertos_20180402_veditado.pdf (accessed on 17 June 2019).

29. IDEA. Spain's National Renewable Energy Action Plan (2011–2020), Plan de Energías Renovables, Instituto para la Diversificación y Ahorro Energético (IDAE). 2010. Available online: http://www.minetad.gob.es/energia/desarrollo/EnergiaRenovable/Paginas/paner.aspx (accessed on 17 June 2019).

30. Wernet, G.; Bauer, C.; Steubing, B.; Reinhard, J.; Moreno-Ruiz, E.; Weidema, B. The ecoinvent database version 3 (part I): Overview and methodology. *Int. J. Life Cycle Assess.* **2016**, *21*, 1218–1230. [CrossRef]

31. European Commission. *Characterisation Factors of the ILCD Recommended Life Cycle Impact Assessment Methods: Database and Supporting Information*, 1st ed.; EUR 25167 2012; European Commission: Brussels, Belgium, 2012.

32. Cole, W.J.; Frew, B.A.; Gagnon, P.J.; Richards, J.; Sun, Y.; Margolis, R.M.; Woodhouse, M.A. *SunShot 2030 for Photovoltaics (PV): Envisioning a Low-Cost PV Future*; U.S. National Renewable Energy Laboratory: Golden, CO, USA, 2017; pp. 1–69.

33. Ram, M.; Child, M.; Aghahosseini, A.; Bogdanov, D.; Lohrmann, A.; Breyer, C. A comparative analysis of electricity generation costs from renewable, fossil fuel and nuclear sources in G20 countries for the period 2015–2030. *J. Clean. Prod.* **2018**, *199*, 687–704. [CrossRef]

34. IRENA. *Fugure of Wind: Technology, Grid Integration and Socio-Economic Aspects*; IRENA: Abu Dhabi, UAE, 2019; pp. 1–88.

35. IRENA. *Future of Solar Photovoltaic: Deployment, Investment, Technology, Grid Integration and Socio-Economic Aspects*; IRENA: Abu Dhabi, UAE, 2019; ISBN 9789292601560.

36. Martikainen, J.P. Response to 'A comparative analysis of electricity generation costs from renewable, fossil fuel and nuclear sources in G20 countries for the period 2015–2030'. *J. Clean. Prod.* **2019**, *208*, 142–143. [CrossRef]

37. *EIA Annual Energy Outlook 2020, with Projections to 2050*; U.S. Energy Information Administration, U.S. Department of Energy: Washington, DC, USA, 2020.

38. Rutovitz, J.; Usher, J. *Methodology for Calculating Energy Sector Jobs*; Prepared for Greenpeace Int. by Institute for Sustainable Future; University Technology Sydney: Sydney, Australia, 2010.

39. Rutovitz, J.; Harris, S. *Calculating Global Energy Sector Jobs: 2012 Methodology*; Institute for Sustainable Future, University Technology Sydney: Sydney, Australia, 2012.

40. del Río, P.; Gual, M.A. An integrated assessment of the feed-in tariff system in Spain. *Energy Policy* **2007**, *35*, 994–1012. [CrossRef]

41. del Río González, P. Ten years of renewable electricity policies in Spain: An analysis of successive feed-in tariff reforms. *Energy Policy* **2008**, *36*, 2917–2929. [CrossRef]

42. San Miguel, G.; Corona, B. Economic viability of concentrated solar power under different regulatory frameworks in Spain. *Renew. Sustain. Energy Rev.* **2018**, *91*, 205–218. [CrossRef]

43. del Río, P.; Mir-Artigues, P. Combinations of support instruments for renewable electricity in Europe: A review. *Renew. Sustain. Energy Rev.* **2014**, *40*, 287–295. [CrossRef]

Article

Combining Biomass Gasification and Solid Oxid Fuel Cell for Heat and Power Generation: An Early-Stage Life Cycle Assessment

Christian Moretti [1,*], Blanca Corona [1], Viola Rühlin [1], Thomas Götz [2], Martin Junginger [1], Thomas Brunner [3], Ingwald Obernberger [3] and Li Shen [1]

[1] Copernicus Institute of Sustainable Development, Utrecht University, 3584 CB Utrecht, The Netherlands; b.c.coronabellostas@uu.nl (B.C.); Viola.ruehlin@bluewin.ch (V.R.); H.M.Junginger@uu.nl (M.J.); l.shen@uu.nl (L.S.)

[2] Division Energy, Transport and Climate Policy, Wuppertal Institute for Climate, Environment and Energy, D-42103 Wuppertal, Germany; thomas.goetz@wupperinst.org

[3] BIOS BIOENERGIESYSTEME GmbH., A-8020 Graz, Austria; brunner@bios-bioenergy.at (T.B.); obernberger@bios-bioenergy.at (I.O.)

* Correspondence: c.moretti@uu.nl

Received: 6 May 2020; Accepted: 27 May 2020; Published: 1 June 2020

Abstract: Biomass-fueled combined heat and power systems (CHPs) can potentially offer environmental benefits compared to conventional separate production technologies. This study presents the first environmental life cycle assessment (LCA) of a novel high-efficiency bio-based power (HBP) technology, which combines biomass gasification with a 199 kW solid oxide fuel cell (SOFC) to produce heat and electricity. The aim is to identify the main sources of environmental impacts and to assess the potential environmental performance compared to benchmark technologies. The use of various biomass fuels and alternative allocation methods were scrutinized. The LCA results reveal that most of the environmental impacts of the energy supplied with the HBP technology are caused by the production of the biomass fuel. This contribution is higher for pelletized than for chipped biomass. Overall, HBP technology shows better environmental performance than heat from natural gas and electricity from the German/European grid. When comparing the HBP technology with the biomass-fueled ORC technology, the former offers significant benefits in terms of particulate matter (about 22 times lower), photochemical ozone formation (11 times lower), acidification (8 times lower) and terrestrial eutrophication (about 26 times lower). The environmental performance was not affected by the allocation parameter (exergy or economic) used. However, the tested substitution approaches showed to be inadequate to model multiple environmental impacts of CHP plants under the investigated context and goal.

Keywords: CHP; biomass; LCA; gasification; SOFC; allocation; multifunctionality

---

1. Introduction

Compared to separate production of heat and electricity from fossil fuels, combined heat and power systems (CHPs) can potentially allow for significant reductions of climate change impact [1,2]. In Europe, coupling heat and electricity generation from renewable sources is also one of the most cost-effective decarbonization strategies [3–5]. In particular, solid biomass has attracted increasing interest by policymakers and investors especially due to the high availability of local biomass from forests and wood processing industries in some regions [6]. The environmental performance of biomass-fueled CHPs depends not only on the type of technology but also on the type of biomass, its supply chain, and the environmental impact categories in focus [7,8].

Mature CHP technologies using solid biomass as fuels have often shown restricted fuel flexibility, limited electric efficiencies, and high particulate matter emissions [9]. To overcome these three limitations, a novel technology was developed during the H2020 HiEff-BioPower project [10]. This novel technology (see Figure 1) is based on a fixed-bed updraft gasifier coupled with a novel primary gas treatment zone, a novel gas cleaning unit (GCU), and a solid oxide fuel cell (SOFC). Its current technological readiness level is between 4 and 5 (based on the definition adopted by the European Commission [11]). The biomass fuel is converted into product gas in the gasifier. Syngas derived from biomass (e.g., wood chips) contains HCl, H2S, and tars [12] making it not suitable for direct utilization in fuel cells [13], which require purified gaseous fuels. Therefore, the syngas from the gasifier is first pre-treated in a primary gas treatment unit (first tar reforming step) and then purified in the GCU. The GCU is one of the key innovations of this technology. It combines the use of ceramic filter candles and sorbents [10]. Syngas cleaning is processed in five steps: primary tar reforming, high-temperature particle filtration, HCl sorption (after cooling the product gas), H2S removal by sorbents and tar reforming (after reheating). After re-heating, the product gas is then fed into the SOFC unit to generate electricity. The off-gases from the SOFC unit are then burnt in a catalytic afterburner to recover heat. Most biomass CHPs are suited for medium and large-scale plants (1–100 MWel). The HBP is available also in small size (about 200 kW of electricity output) [9]. Among biomass technologies of this size, one of the main competitors is the organic Rankine cycle (ORC) [14].

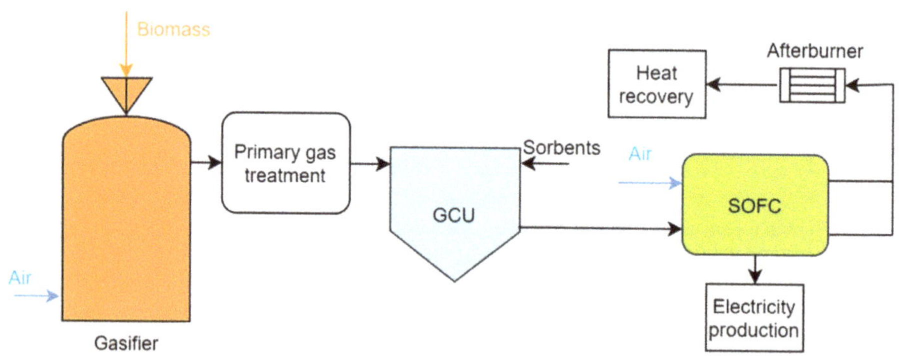

Figure 1. Concept of high-efficiency bio-based power (HBP) technology. GCU = Gas cleaning unit. SOFC = Solid oxide fuel cell.

At this stage of the Hieff-BioPower project, the assessment of the environmental impacts of the current design configuration can help with minimizing impacts at an early stage of the technical HBP development. In particular, the literature reports a few life cycle assessments (LCAs) of heat and power from SOFC-based CHPs and several ones of CHPs involving biomass gasification processes but no one on their combination (in October 2019, from Scopus database searching in TITLE-ABS-KEY). These studies provided the following main findings: (1) the investigated CHPs present lower impact in terms of climate change compared to conventional technologies [2,15] and (2) the biomass fuel production has the highest contribution to total life cycle impacts [16,17]. These studies also highlighted several methodological uncertainties of LCAs that can lead to significantly different results. Such uncertainties are mainly linked to the multifunctional nature of the CHPs. A CHP is a system producing two products, heat and electricity. Depending on the goal of the LCA, it may be necessary to apportion the overall impact of the system to each of the co-products. Finding the right criterion for the allocation of impacts to each co-product is generally understood as a multifunctionality problem [18]. When a multifunctionality issue is encountered, the practitioner has to properly select the functional units and allocation methods [19,20]. The selected criterion could affect the outcome of the LCA significantly and, for this reason, this selection is broadly discussed in the literature [21,22].

The environmental LCA presented in this study has a twofold aim: (1) to identify the main sources of the environmental impact of this new technology and (2) to assess its ecological competitiveness compared to the separate production of heat and electricity and one of its main competitors, i.e., Organic Rankine Cycles (ORC). Moreover, this case study is used to analytically discuss the influence of the allocation method in the LCA results for CHP plants and provide methodological recommendations for better allocation practices.

2. Materials and Methods

2.1. Goal and Scope Definition

The LCA has been conducted according to ISO 14040:2006 and ISO 14044:2006 [18,23]. The intended audience of this LCA consists of technology developers, researchers involved in the field of bioenergy and LCA practitioners. An attributional LCA (ALCA) approach is followed since the goal of this study is to identify the activities within HBP causing the highest contribution to the environmental impacts, and not the consequences of changes in these activities [21,24].

Two technologies are considered for environmental comparison: (1) a combination of the electricity mix (EMIX) from the German national grid plus heat provided from a natural gas boiler (NG) and (2) biomass-based organic Rankine cycle (ORC) CHP.

As the ORC CHP has a different heat to electricity ratio compared to the HBP, the definition of two functional units was preferred to the definition of a single functional unit with a fixed heat/electricity ratio. Hence, two functional units were defined as follows: 1 kWh of electricity or 1 MJ of heat.

The HBP technology finds one of the main strengths in its fuel flexibility [10] since it can operate with various biomass feedstocks in the forms of chips or pellets. To explore the effect of different feedstocks on the environmental impacts of the HBP CHP technology, this study explored the use of three different types of biomass fuels: wood chips, wood pellets, and *Miscanthus* pellets. The operation with wood chips was considered as the baseline scenario (WC), while the operation with wood pellets (WP) and the operation with *Miscanthus* pellets (MP) as alternative scenarios. The baseline scenario with wood chips was also used for comparison with the competing technologies, i.e., the ORC technology (fueled with wood chips as well) and the combination of grid electricity plus natural gas boiler. Additionally, this last competing option was also compared to the WP and MP alternative scenarios.

Figure 2 shows the process diagram of the HBP product system. The system boundaries follow a cradle-to-gate approach. As shown in Figure 2, all the life cycle stages from the extraction of the raw materials to the final dismantling and waste treatment are included. The final distribution and consumption of the products, i.e., heat and electricity, are not included in the LCA. After the power plant is dismantled and parts are recycled, the use of the recycled materials is outside of the system boundaries. Biomass transport stages from the forest to the processing plant and from the processing plant to the HBP plant were included in the study. The transportation of plant components (e.g., the gasifier) from the production site to the power plant location and the construction activities of the plant were not included in the analysis. The exclusion of these activities was based on their expected minor contribution to the total environmental impacts, as also found in similar studies, for example [16].

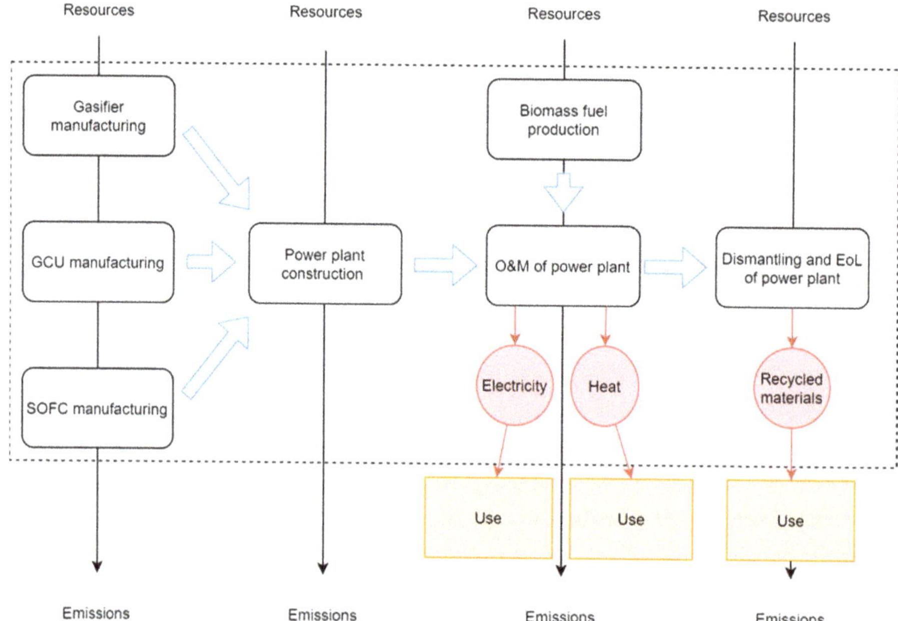

Figure 2. Flowchart of the HBP product system, including system boundaries (dashed lines).

The temporal scope of the study is placed in the near future (the next 5–10 years) when the HBP technology should be commercialized. The HBP is assumed to be installed in Germany, being the country with the maximum potential sales for the HBP technology in Europe [25]. Nevertheless, some components for the HBP (e.g., the gasifier) might also be manufactured outside Germany (in other EU countries).

Seven mid-point impact categories were selected and the adopted impact assessment models for each impact category were selected following the ILCD recommendations [26] (see Table 1). Climate change (CC) and depletion of mineral, fossil, and renewable resources (MFRD) were chosen because they are considered top priorities in the current societal and political challenges [27]. Particulate matter (PM) and photochemical ozone formation (POF) are selected because of their relevance to the energy sector [28]. Acidification (AC), Terrestrial eutrophication (TE), and Water resource depletion (WRD) were selected because of their relevance for agricultural systems, and therefore for biomass production [29].

To assess the robustness of the results, two sensitivity analyses were conducted. As anticipated in the introduction, a comprehensive sensitivity analysis was performed on the allocation choices to explore their influence in the outcome of the LCA (and as recommended by ISO [18]). The second sensitivity analysis was performed to explore parameters that are potentially sensitive for the results and that might environmentally improve or make less attractive the technology in the future.

Table 1. Selected impact categories and models.

Impact Category	Unit	Impact Assessment Models
Climate change (CC)	kg CO_2eq	IPCC 2013 Global Warming Potential 100 years [30]
Particulate matter (PM)	kg PM2.5 eq	Premature death or disability from particulates/respiratory inorganics from [31]
Photochemical ozone formation (POF)	kg NMVOC eq	Potential contribution to photochemical ozone formation for Europe from [32]
Acidification (AC)	molc H+ eq	Accumulated Exceedance (AE) characterizing the change in critical load exceedance of the sensitive area from [33]
Terrestrial eutrophication (TE)	molc N eq	Accumulated Exceedance (AE) characterizing the change in critical load exceedance of the sensitive area from [33]
Water resource depletion (WRD)	m^3 water eq	Freshwater scarcity: Scarcity-adjusted amount of water used from Swiss Ecoscarcity 2006 [34]
Mineral, fossil and renewable resource depletion (MFRD)	kg Sb eq	Depletion of resources based on the scarcity model from [35]

2.2. Life-Cycle Inventory

2.2.1. Unit Processes, Data, and Assumptions

This study assesses the small scale configuration of the HBP technology, which has a nominal electricity output of the SOFC of 199 kW. Its main characteristics during the average lifetime (assumed 18 years) are reported in Table 2.

Table 2. Characteristics of the small scale HBP technology (8000 h of operations per year). Modeled values.

Flow (Unit)	Wood Chips (30 wt.%)	Pellets (from Wood and *Miscanthus*) (5 wt.%)
Biomass fuel (kW)	548.5	570.0
Biomass fuel (kg/h)	164.1	115.7
Gross electric power (kW)	170.5	190.0
Thermal power output (kW)	288.5	292.0
Electrical efficiency gross (%)	31%	33%
Electrical efficiency net (%)	30%	32%
Thermal efficiency (%)	53%	51%
Exergy output as heat (%)	24.6	22.7
Exergy output as electricity (%)	75.4	77.3
Economic output as heat (%)	41.8	39.3
Economic output as electricity (%)	58.2	60.7

For the foreground system, data on the gasifier and the GCU were collected from the industrial partners involved in the H2020 HiEff-BioPower project. For the SOFC, secondary data based on the scientific literature [17] were used due to the unavailability of specific primary data. The background data were largely based on the ecoinvent database (version 3.4). For unavailable data, assumptions were made based on literature (see the following sub-sections for details regarding each phase of the life cycle).

As the system provides two different products (and functional units), it was necessary to determine an allocation key to partition the overall impact to the two functional units. Allocation by physical causality was not applied under the absence of a representative mathematical model (to model the causality relationships) [36–38]. Among the possible remaining allocation methods, the exergy key was chosen because it can represent both quantity and quality of both functional outputs, and is common practice for CHPs (e.g., ecoinvent uses such key [39] and is recommended by RED II [40]). Table 2 shows the biomass input, intermediate performance indicators, and energy outputs in terms of their exergy and economic values. The exergetic outputs expressed in percentage reported in Table 2 represent also the allocation factors used for the baseline calculations. The economic values are based on three years (2015–2017) average prices for medium size industries without VAT, in Germany, retrieved from Eurostat [41]. The prices were 0.079 € per kWh of industrial electricity and 0.0086 € per MJ of industrial heat.

2.2.2. Inventory Data for Chips and Pellets

To model the life cycle of wood chips, the ecoinvent 3.4 dataset "Wood chips, wet, measured as dry mass {CH}|market for|APOS" was used. This dataset includes both wood chips from industrial activities and forest management and represents the average Swiss market (assumed to be a good proxy for Germany). In particular, wood chips from forest management represents an 85% share of the modeled Swiss wood chip market.

For wood pellets, the ecoinvent 3.4 dataset "wood pellet, measured as dry mass {RER}|market for wood pellet|APOS" was used.

For *Miscanthus* pellets, a similar dataset was not available in ecoinvent. Hence, the inventory data from [42] were used together with the best practices reported in [43]. An average dry yield value of 23.5 t *Miscanthus* (85% dry matter) per hectare was used to estimate the land requirements to provide enough fuel for the HBP plant for one year. The planting rate of 16,000 *Miscanthus* per ha was taken from [43]. As *Miscanthus* is a perennial crop, field preparation activities such as herbicide application, harrowing and plantation, occur only during the first year. The lifetime of the crop was assumed to be 18 years [43] and therefore 1/18 of the impact from field preparation activities was apportioned to one year of operation of the HBP plant. Once the *Miscanthus* is collected from the field, it is necessary to transport it to the pelleting plant. The transport distance to the pelleting plant was assumed to be 10 km by tractor [42]. For the chipping of *Miscanthus*, the energy consumption of the chipper and the amount of lubricating oil were retrieved from the ecoinvent 3.4 datasets "Wood chips, wet, measured as dry mass {CH}|wood chips production, hardwood, at sawmill|APOS". For the pelleting of *Miscanthus*, the amounts of electricity, heat, lubricating oil, and water were retrieved from the ecoinvent 3.4 dataset "Wood pellet, measured as dry mass {RER}|wood pellet production|APOS". The transportation of *Miscanthus* pellets to the HBP plant was assumed to occur by truck and with an average distance of 100 km [42].

2.2.3. Inventory Data for the Manufacturing of the Power Plant

The HBP manufacturing consists of three sub-processes: the manufacturing of the gasifier, the manufacturing of the SOFC stack and its balance of plant (BoP), and the manufacturing of the GCU. The data for the manufacturing of the gasifier is based on HBP project data and shown in Table 3.

Table 3. Materials of the gasifier including the primary gas treatment zone.

Material	Amount	Process Dataset
Steel (low alloyed) (kg)	6770	Steel, low-alloyed {RER}|steel production, converter, low-alloyed|APOS
Stainless steel (kg)	585	Steel, chromium steel 18/8, hot rolled {RER}|production|APOS
Iron-nickel-chromium alloy (kg)	220	Iron-nickel-chromium alloy {RER}|production|APOS
Concrete fireproof (kg)	4480	Concrete block {DE}|production|APOS
insulating material (kg)	1220	Glass wool mat {CH}|production|APOS

Concerning the SOFC stack, its production was modeled considering secondary data from scientific literature and, to a lower extent, from ecoinvent database. The literature data was retrieved from studies where the SOFC stacks had a similar power capacity as the HBP technology. The amount of electricity, nickel oxide, solvents, materials for the binder, carbon black, and chromium steel, as well as direct emissions (released during the production of the stack) to the air of carbon dioxide, methyl ethyl ketone, and benzyl alcohol were taken from [17] and adjusted proportionally to the power capacity (factor of 0.793 based on 199 kWe of HBP SOFC versus 250 kWe of SOFC in [17]).

The data for the manufacturing of the anode, cathode, electrolyte, and the required ceramic materials (Lanthanum Strontium Manganite (LSM) and Yittria Stabilised Zirconia (YSZ)) were retrieved from [44].

The other secondary data for the SOFC stack, which were not available in [17,44], were retrieved from the already existing inventory in ecoinvent 3.4 called "Fuel cell, stack solid oxide, 125kW electrical, future {CH}\|production APOS" and multiplied times 1.59 to account for the different size (assumption of linear proportionality of materials to the size as before).

For the production of the SOFC's BoP, data for the inputs of steel and energy were retrieved from [17]. The other data were instead retrieved from ecoinvent 3.4 dataset "Fuel cell, solid oxide, 125 kW electrical, future {CH}\|production\|APOS", which was modified as well by multiplying times 1.59.

The materials for manufacturing the cage of the GCU were assumed to be similar to the ones of the cage of the external reformer of the SOFC provided in ecoinvent 3.4. The 96 filter candles which are present in the GCU system at the beginning of the operation were included within the manufacturing stage. These candles are made from calcium-magnesium-silicate high-temperature fiberglass. The processes "Calcium borates {GLO}\|market for\|APOS", "Magnesium {GLO}\|market for\|APOS" and "Silica sand {DE}\|production\|APOS" from ecoinvent 3.4 were used as a proxy for CaMgO4Si. It was further assumed that 1.1 kg of material input would generate 1 kg of filter candles. The mass of each candle was derived from the technical sheet of the candles [45].

2.2.4. Inventory Data for Operation and Maintenance

The system operation includes all the material and energy inputs needed to operate the plant during one year of service (e.g., gas cleaning sorbents, water), waste outputs (e.g., ash which needs to be disposed of) and direct emissions to the environment (e.g., pollutant gas released to air).

The resulting direct emissions to air from the HBP are summarised in Table 4. Data for such emissions were only available for wood chips and wood pellets. The emissions from the operation with *Miscanthus* pellets were assumed to be the same as for wood pellets. Data on the ash formation (grate ash and fly ash) was retrieved from [14].

Table 4. Direct emissions (mg) to air per MJ of overall energy output (heat and electricity). OGC = organic gaseous compounds, TSP = total suspended particle, NOX = nitrogen oxides. Maximum values shown in the table were used in the Life Cycle Inventory.

Fuel	CO	OGC	TSP	NO$_X$
Wood chips	<20	<0.01	<0.01	<0.01
Pellets	<20	<0.01	<0.01	<0.01

The operation of the gasifier needs 2.36 kg of natural gas for start-up operations and about 80.0 t of tap water per year for gasification air humidification (based on simulations from project data). According to measurements performed downstream the primary gas treatment zone, i.e., at GCU inlet, the syngas composition during utilization of wood chips is as follows (in volume percentage): 15.4% CO, 10.6% CO_2, 1.8% CH_4, 8.3% H_2, 21.8% H_2O, 41.1% N_2. During the multiple tests performed, such a composition showed to be stable. After the primary treatment unit, the syngas typically shows contaminant concentrations in the range of 30 ppm for sulfur and 20 ppm for chlorine (on wet basis) when wood chips are used as fuel. The tar concentration at the inlet of the GCU was lower than 2.0 g/Nm^3 on a dry basis and the particulate matter contents (TSP) of about 200 g/Nm^3 on a dry basis were determined.

For the operation of the GCU, about 1.2 t of zinc oxides per year are needed for H2S removal. The GCU also requires 1200 Nm^3 of Nitrogen per year for the cleaning of the filter elements. One year of operation of the GCU requires also 4800 kg of dolomite mixed with 900 kg of sodium bicarbonate (from ecoinvent 3.4, Soda ash, dense {GLO}|market for|APOS) as coating materials respectively for Cl-sorption. The GCU has been designed to feed the SOFC with a product gas containing less than 5 ppm of chlorine, less than 1 ppm of sulfur, and less than 100 ppm of particulate matter (TPS < 0.1 mg/Nm^3, on wet basis). Since the composition of the syngas is expected to be stable (confirmed also by the first test runs), the uncertainty about the simulated electric power output of the SOFC is expected to be very low.

The maintenance stage includes all the components which are replaced during the lifetime of the HBP plant. The SOFC stack and the GCU have a shorter lifetime than the average lifetime of the HBP CHP plant. Since the SOFC stack currently investigated for the HBP technology has an estimated lifetime of 5 years, the production of 1/5 extra SOFC stack per year was added to the maintenance stage. The GCU used for the HBP technology has a lifetime of 10 years, therefore, the production of 1/10 extra GCU per year was added to the maintenance stage. All other maintenance inputs (steel components and deionized water consumption for start-ups) for the SOFC were retrieved from [17] and scaled for the capacity of the SOFC under investigation. For the filter candles, an average of 30% of candles is estimated to be replaced each year of operation of the HBP technology. Therefore, the production of 29 extra candles per year was included in the maintenance stage.

2.2.5. Inventory Data for End-of-Life Disposal

The main material employed in the components of the HBP is steel and can be recycled at the end of the life of each component. Based on the amount of steel present in the components (and their replacements), it was assumed that about 1900 kg of steel are recycled per average year of operation. The model included the energy for pressing and crushing the steel crap (based on [46]), a recycling efficiency (referred to as RRE in Equation (1)) of 88% [47], and a transportation distance of 100 km from [46]. Such transportation was assumed to occur mainly by 16–32 t lorries Euro 3 [46].

A recycling process is a typical example of a multifunctional process fulfilling two functions i.e., the treatment of waste and the production of a recycled product. Based on our goal, the modeling approach (i.e., attributional), and the recommendations by ISO 14044:2006, mass allocation was applied. This selection is based on the fact that ISO 14044:2006 prioritizes allocating by a physical property for open-loop recycling over the economic value or number of uses (among ISO third level allocations i.e., by other relationships). Additionally, system expansion cannot be applied since we want to isolate the first function (first use of the material which led to its treatment) from the second function (next use or cycle of the material). The impacts arising from transportation (E_T), recycling (E_{RC}), and the extraction and processing of the primary material (E_v) were therefore allocated by mass between this life cycle and the following one (see Equation (1) expressing the allocated impact to our functions). The resulting mass allocation factor (1/(1 + RRE)) was 53% (1/1.88). The second part of Equation (1) related to the virgin material takes into account the fact that the primary production was already accounted entirely in the manufacturing phase, and therefore the corresponding burdens (e.g., extraction of raw material) that belong to the following life cycle should be subtracted.

$$E_{\text{steel disposal}} = \text{RRE}(E_T + E_{RC}) \frac{1}{1 + \text{RRE}} - \frac{\text{RRE}}{1 + \text{RRE}} E_v \tag{1}$$

There are some precious metals (e.g., used as catalytic materials) used in the power plant that, depending on the recovery efficiency and initial concentration, might be economically convenient to recover, though e.g., hydrometallurgical treatment [48]. Nevertheless, such specific recovery processes were not modeled because of the unavailability of Life Cycle Inventory data. Materials other than the steel used in the power plant components consist of hazardous waste (24 kg) and inert waste (10 t per year in the chips scenario, and 20 t per year for the pellets scenarios). The treatment of the hazardous waste was modeled through the ecoinvent dataset "Hazardous waste, for underground deposit {DE}|treatment of hazardous waste, underground deposit|APOS". The inert waste consists mainly of materials for sorbents and was modeled through the ecoinvent 3.4 dataset "Inert waste, for final disposal {CH}|market for inert waste, for final disposal|APOS".

2.2.6. Inventory Data for the Competing Technologies

For the comparative analysis, the ecoinvent 3.4 datasets "1 MJ Heat, district or industrial, other than natural gas {CH}|heat and power co-generation, wood chips, 2000 kW, state-of-the-art 2014|APOS" and "1 kWh Electricity, high voltage {CH}|heat and power co-generation, wood chips, 2000 kW, state-of-the-art 2014|APOS" were used for the ORC. This dataset represents a state of the art ORC co-generation plant equipped with an electrostatic precipitator for particulate emission reduction and includes the infrastructure. For the separate production of heat and electricity, the ecoinvent 3.4 datasets "1 MJ Heat, central or small-scale, natural gas {CH}|heat production, natural gas, at boiler condensing modulating < 100kW|APOS" and "1 kWh Electricity, medium voltage {DE}|market for|APOS" were used.

Following the description provided in ecoinvent 3.4 for the ORC ecoinvent dataset, the capacity of the ORC plant is 1000 kW thermal, and 200 kW electric (similar to the electric output of the HBP technology). This information was used to estimate the exergy allocation factor of 46% for heat (assumed district heating provided at 90 °C as for HBP) and 54% for electricity. Based on 2015–2017 average prices for Germany, the economic allocation shares for ORC would be 66% for heat and 34% for electricity. As the total power input (as wood chips) is 2000 kW, this ORC plant has an overall energy efficiency of 60%, i.e., 10% electrical efficiency plus 50% thermal efficiency.

3. Results

3.1. Environmental Impact of the HBP Technology

3.1.1. Baseline (Wood Chips)

Figure 3 shows the breakdown of the impact of the HBP technology for the seven investigated impact categories (see Appendix A for absolute values).

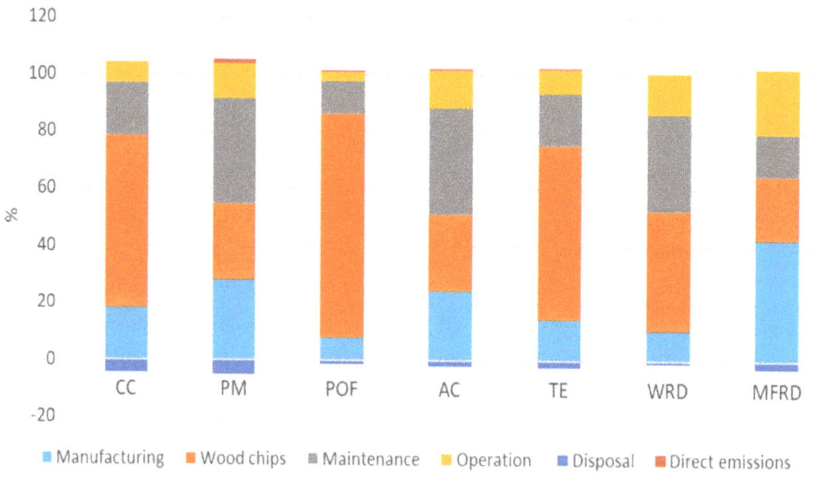

Figure 3. Main contributions to the cradle-to-grave environmental impact of producing heat and electricity with the HBP technology using wood chips as biomass fuel. The presented breakdown is valid for both functional units. CC = Climate change, PM = Particulate matter, POF = Photochemical ozone formation, AC = Acidification, TE = Terrestrial eutrophication, WRD = Water resource depletion, MFRD = Mineral, fossil, and renewable resource depletion.

The main contributions to the cradle to grave environmental impact are the wood chips used, followed by the maintenance phase and manufacturing phase. The impact of wood chips is made of two components i.e., their transportation and their production. The impact of transporting wood chips (based on the Swiss supply chain assumed by the dataset retrieved from ecoinvent) represents 18–27% of the impact of wood chips for all impact categories, except for photochemical ozone formation (9%) and water depletion (2%). In all impact categories, except water depletion, the impact of the production of wood chips is mainly caused by the production and combustion of diesel and petrol (60–80%) used in power sawing machines, skidders, and chippers. The production of the lubricants used in the three processes mentioned above causes about 2–10% of the impact of wood chips production in all impact categories except for water resource depletion for which it represents 80% of the impact. The water depletion impact of wood chips is mainly due to the fraction of vegetable oils used for lubricating the chains during power sawing activities (in absolute terms, this impact is quite low, see figures in Section 3.1.2 and Section 3.2.2.

The main impact of the manufacturing stage is due to the production of the SOFC system, which contributes to 63–100% of the impacts in this stage (depending on the category). Within the SOFC system, the production of the SOFC stack and the inverter are the main sources of impact. This is mainly due to the large electricity consumption during the manufacturing of the stack (as also highlighted by Rillo et al. [17]) and the manufacturing of chromium steel (mainly caused by the production of ferrochrome [17]).

Concerning the maintenance impacts, the maintenance of the SOFC system contributes to 63–100% of the impact depending on the impact category. The major contributor (95–99%) to the impact of the maintenance of the SOFC system is the replacement of the SOFC stack, which requires the production of a new SOFC stack every five years of operations.

The operation phase is dominated by the operation of the GCU (mainly zinc oxide used and sodium bicarbonate) expect for water depletion whose impact is mainly caused by the water used for the operation of the gasifier. The contribution of direct emissions is negligible in all impact categories.

The particulate matter caused by the operation of the HBP technology was only 1% of the total particulate matter impact.

3.1.2. Alternative Scenarios (Wood and *Miscanthus* Pellets)

Figure 4 shows the environmental impact of the baseline scenario in comparison to the alternative scenarios.

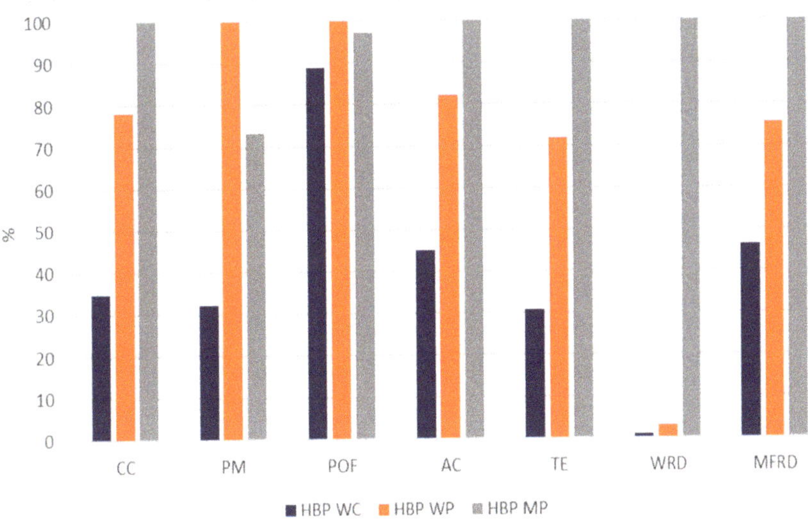

Figure 4. HBP technology fueled with various biomass fuels (the same graph applies to both 1 MJ heat or 1 kWh electricity). Values are normalized taking as 100% the impacts of the most impacting scenario. CC = Climate change, PM = Particulate matter, POF = Photochemical ozone formation, AC = Acidification, TE = Terrestrial eutrophication, WRD = Water resource depletion, MFRD = Mineral, fossil, and renewable resource depletion. WC = wood chips, WP = wood pellets, MP = *Miscanthus* pellets.

The results show that, in all impact categories, the total life cycle impact is the lowest for the operation with wood chips compared to the other two biomass scenarios (wood pellets and *Miscanthus* pellets). Since the inventories for manufacturing and maintenance are the same, the main difference between the three scenarios is the production of the biomass fuel (wood chips have lower environmental impacts than the two pellets).

The impact of the WC scenario is between 10% and 70% lower than for the WP scenario (with the highest impact difference for water depletion and particulate matter). For water depletion, the impact of wood pellets is almost entirely caused by the electricity consumption of the pellet factory. For particulate matter, the shaving process accounts for about 54% of the impact of producing wood pellets. Shaving is, therefore, the main cause of the significantly higher particulate matter impact in the production of wood pellets compared to wood chips. The impact of shaving is mainly caused by its drying process (87%), which leads to high particulate emissions due to the combustion of industrial wood.

Except for particulate matter and photochemical oxidant formation, the *Miscanthus* scenario presents higher environmental impacts than the wood pellets scenario. The characterized results indicate between 18% and 28% lower impacts for the wood pellets scenario than for the *Miscanthus* scenario in the categories of acidification, climate change, resource depletion (mineral, fossil, and renewables), and terrestrial eutrophication. The particulate matter impact is lower (−27%) in the case of *Miscanthus* pellets because the shaving process, which was the main source of impact for wood pellets, is not used to produce *Miscanthus* pellets.

The difference in impact is even higher for the water depletion category, which scores 97% lower in the wood pellets scenario than in the *Miscanthus* pellets scenario. The irrigation needed during its cultivation is the main cause of the significantly higher water depletion in the scenario with *Miscanthus* pellets (see Figure 5). Other activities that are an important source of impacts for *Miscanthus* pellets are the electricity for pelleting, the diesel burnt during the harvesting stage and the emissions caused by fertilizing (see Figure 5 for the single contributions in each impact category).

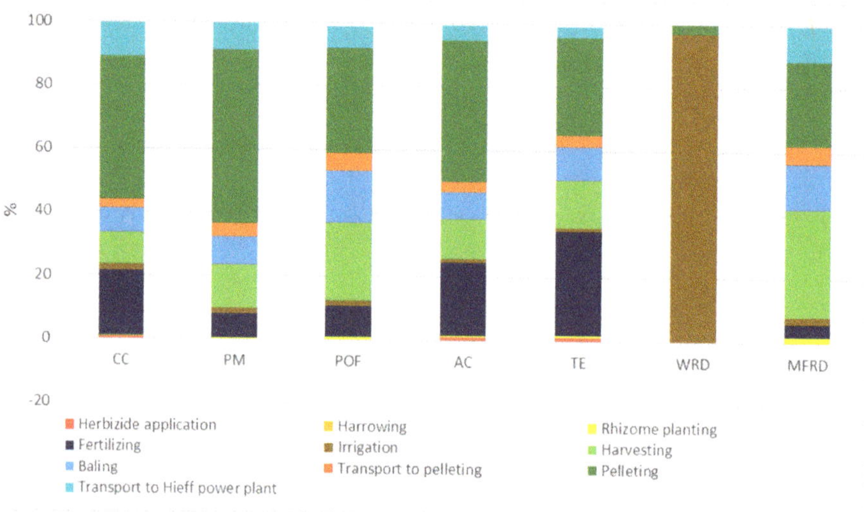

Figure 5. Main contributions to the environmental impact of *Miscanthus* pellets supplied to the HBP CHP plant. CC = Climate change, PM = Particulate matter, POF = Photochemical ozone formation, AC = Acidification, TE = Terrestrial eutrophication, WRD = Water resource depletion, MFRD = Mineral, fossil, and renewable resource depletion.

Similar to the baseline case of wood chips, direct emissions have a negligible impact on the operation with wood pellets and *Miscanthus* pellets. This aspect is particularly important in the case of biomass technologies installed in heavily populated areas.

As *Miscanthus* is an energy crop, it is important to assess the impacts due to land use. As for the other impact categories, the selection of the method was based on ILCD recommendations [26]. Accordingly, the carbon deficit caused by land use was assessed using the Soil Organic Matter model of [49]. This model accounts for the changes in soil quality caused by the occupation and transformation of the land. Land occupation generates changes in soil quality which depend on the amount of area occupied and the duration of such an occupation. Land transformation generates changes in soil quality which depend on the extent of changes in land properties and the area affected. In this model, the deficits in soil organic matter content are assessed and expressed by an indicator whose unit is kilograms of carbon deficit. These deficits are caused by the effects of agricultural practices on degradation rates. The changes can also be additions of soil organic matter. For example, these additions can be caused by the application of manure or crop residues. It should be observed that this modelling of land use impacts does not account for the counterfactual effects caused by land use changes modelled in consequential LCAs of bioenergy.

The production of 1 MJ of heat using *Miscanthus* pellets generates a 0.86 kg C deficit. Such an impact is much higher than for 1 MJ of heat generated using wood chips (0.12 kg C deficit) and using wood pellets (0.13 kg C deficit). The reason is that *Miscanthus* is an energy crop. Hence, differently from the feedstock for wood chips and pellets, it requires dedicated cultivation.

3.2. Benchmarking with Competing Technologies

3.2.1. Comparison with ORC Technology (Both Fueled with Wood Chips)

Figure 6 shows the comparison between HBP technology and ORC technology both fueled with wood chips.

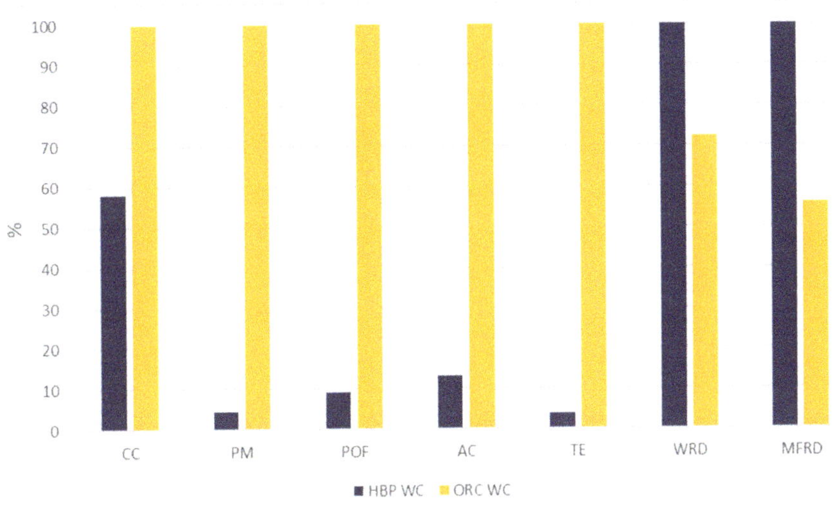

Figure 6. Comparison of HBP technology with ORC technology for 1 MJ of heat. The graph for 1 kWh electricity shows some minor differences due to a slightly different Carnot factor assumed for ORC technology. Values are normalized with respect to the impacts of the most impacting scenario. CC = Climate change, PM = Particulate matter, POF = Photochemical ozone formation, AC = Acidification, TE = Terrestrial eutrophication, WRD = Water resource depletion, MFRD = Mineral, fossil, and renewable resource depletion. WC = wood chips.

Compared to the same amount of heat produced by the ORC technology, the heat co-generated by the HBP technology shows lower environmental impact in terms of climate change (−42%) and much lower impact (−87%/−96%) in terms of particulate matter, photochemical ozone formation, acidification and terrestrial eutrophication. These differences can be explained by two main advantages of the HBP technology: (1) the HBP has higher energy and exergy efficiencies, and therefore less biomass is needed for producing the same amount of energy and exergy as outputs and (2) HBP avoids the external combustion of biomass occurring in the ORCs, and therefore releases less particulate emissions (the particulate matter impact of the ORC technology is for 97% caused by direct emissions of particulates). Although HBP technology has the same thermal efficiency as the ORC technology, its electric efficiency is three times higher. On the other hand, the HBP technology shows higher water depletion (+38%) and resource depletion (+79%). For water depletion, the high impact is caused by the replacements of the SOFC stack. For depletion of resources (minerals, fossil, and renewables), the main cause can be found in the production of the SOFC stacks. All these components are not present in the case of an ORC.

Similar results were obtained when comparing electricity production from HBP and ORC. The HBP shows lower impacts for climate change (−45%), particulate matter (−96%), photochemical ozone formation (−91%), acidification (−88%) and terrestrial eutrophication (−96%). On the other hand, the HBP technology shows an increased impact in terms of water (+31%) and depletion of resources (MFRD) +70%.

3.2.2. Comparison with Conventional Production of Heat and Electricity

Figures 7 and 8 show the comparison between the HBP technology operating with the three investigated biomass fuels and conventional separate productions of heat and electricity.

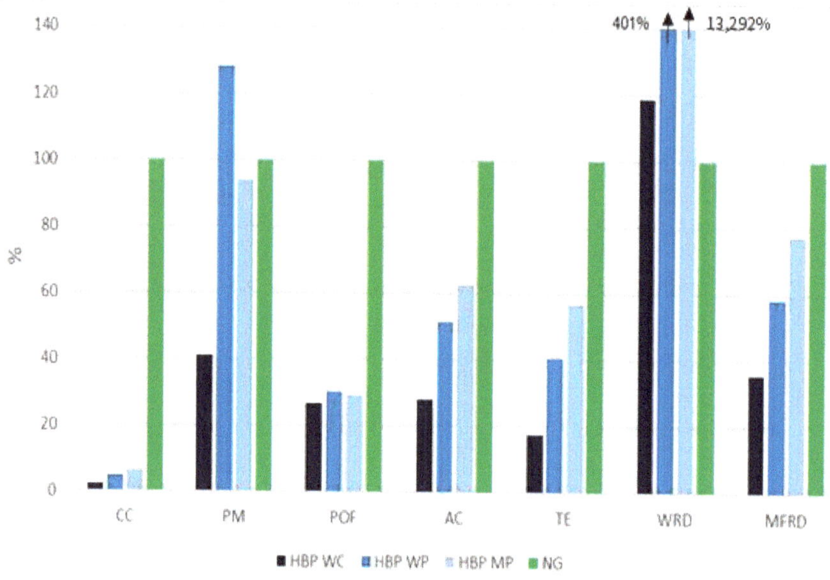

Figure 7. Comparison of HBP technology with competing technologies (for 1 MJ heat). Values are normalized taking as 100% the impacts of the natural gas boiler. CC = Climate change, PM = Particulate matter, POF = Photochemical ozone formation, AC = Acidification, TE = Terrestrial eutrophication, WRD = Water resource depletion, MFRD = Mineral, fossil, and renewable resource depletion. WC = wood chips, WP = wood pellets, MP = *Miscanthus* pellets.

The heat co-generated by the HBP technology shows a lower environmental impact compared to the heat produced by a condensing boiler burning natural gas. Even considering the least preferred fuel scenario in each impact category, the impact differences are at least −94% in terms of climate change, −70% in photochemical ozone formation, −37% in acidification, −43% in terrestrial eutrophication and −22% in depletion of resources. In particular, the significant difference in climate change is mainly generated by the biogenic carbon dioxide emissions (which are assumed to be carbon neutral) instead of fossil ones.

On the other hand, the HBP technology causes +28% impacts in particulate matter in the WP scenario (caused by the high particulate matter released when producing wood pellets) and significantly higher water depletion for the MP scenario (+13,000%), due to the water used for irrigation in the cultivation of *Miscanthus* (the only scenario with irrigation). When wood chips are fed instead of pellets, the HBP shows a much lower impact in terms of particulate matter (−59%) but still a relatively higher impact in water depletion (+19%), due to indirect water consumption in different life cycle activities.

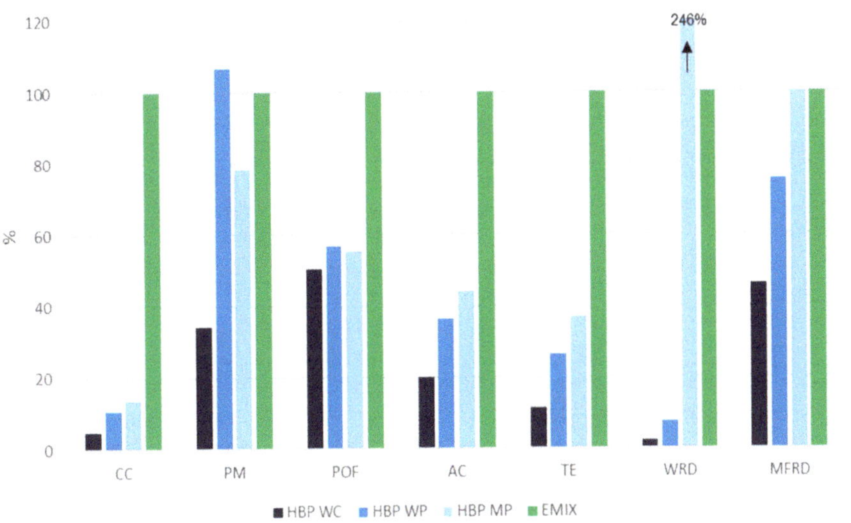

Figure 8. Comparison of HBP technology with competing technologies (for 1 kWh electricity). Values are normalized taking as 100% the impacts of the German electricity mix. CC = Climate change, PM = Particulate matter, POF = Photochemical ozone formation, AC = Acidification, TE = Terrestrial eutrophication, WRD = Water resource depletion, MFRD = Mineral, fossil, and renewable resource depletion. WC = wood chips, WP = wood pellets, MP = *Miscanthus* pellets.

The electricity co-generated by HBP technology shows a lower environmental impact compared to electricity produced by the German electricity mix (EMIX). In particular, even considering the worst fuel scenario, the differences of impact are at least −86% for climate change, −43% for photochemical ozone formation, −56% for acidification and −63% for terrestrial eutrophication. Nevertheless, the HBP using wood pellets as fuel can lead to an increase in particulate matter (+7%). The HBP using *Miscanthus pellets* has much higher water resource depletion (+146%; caused by irrigation of *Miscanthus*) than the electricity from the grid mix. When operating with wood chips, the HBP shows a much lower impact than the EMIX, leading for example to −66% impacts in particulate matter, −98% in water depletion and −54% in depletion of resources (MFRD).

3.2.3. Comparing with Other LCAs of SOFC CHPs

In the literature, 8 LCAs of SOFC CHPs have been conducted along with a review of LCAs on SOFC systems. In most of these LCAs, the fuels used in the SOFC CHPs assessed were natural gas and biogas and the capacity of the SOFC was only a few kilowatts (1–20 kW) of electricity. LCAs on SOFC CHPs of larger capacity (comparable to the one of the HBP) were conducted by [50–52]. Our results for the climate change impact of the HBP technology (0.03–0.09 kgCO$_2$eq/kWhel depending on the fuel considered) indicate considerably lower impacts than for the SOFC CHPs assessed by these LCAs.

These lower impacts are especially found for the SOFC CHPs using natural gas as fuel because of the avoidance of direct emissions of fossil CO$_2$ allowed by the HBP which is fueled with a biofuel instead of fossil fuel. In particular, among the LCAs of SOFC CHPs whose size is comparable to the HBP and operating with natural gas, Strazza et al. [50] assessed a 230 kWel SOFC CHP with electric efficiency of 53.4%. The resulting impact was 0.47 kgCO$_2$eq per kWh of electricity, which is at least 5 times higher than for the HBP. An older study [52] assessing a 125 kWel SOFC CHP operating with natural gas, calculated an impact of 0.9–1.0 kgCO$_2$eq per kWh of electricity, which is at least 10 times higher than for the HBP. Staffell et al. [44] assessed a 1 kWel micro-SOFC CHP fueled with natural gas and calculated an impact of 0.32–0.37 kgCO$_2$eq/kWhel, which is a least 3–4 times higher than for the HBP.

The impact of the HBP is also at least 44% lower than for SOFC CHPs operating with biogas. In this case, the main reason can be found in the different fuel production processes and composition of the fuel used and consequent different composition of the direct emissions (e.g., methane emissions released with the exhaust gases). In particular, Strazza et al. [50] calculated an impact of 0.16 kgCO$_2$eq for a 230 kWel SOFC CHP with 52.2% electric efficiency operating with biogas from sewage sludge. For the same type of system but with a capacity of 125 kWel, Sadhukhan [51] calculated an impact of about 0.19 kgCO$_2$eq per kWhel. Concerning this last figure, Sandhukhan used a different functional unit (1 ton of sewage sludge processed through anaerobic digestion) and we derived it by applying exergy allocation on the energy outputs.

Similarly to the HBP, for multiple impact categories, Strazza et al. found that the impact of this system, independently on the fuel considered (natural gas or biogas) was dominated by the production of the fuel. The only exception was the climate change impact of the operation with natural gas, whose impact was mainly caused by the operation phase (mainly direct emissions of fossil CO$_2$ of the system).

3.3. Alternative Methods for Solving Multifunctionality

Exergy allocation was used to partition the total environmental impact between heat and electricity, as explained in Section 2.2.1. In the literature, the two most applied alternative approaches to address the multifunctionality of SOFC CHPs are system expansion (enlargement) and economic allocation [20]. The first approach was only applied in studies where it was not necessary to differentiate between the impacts of heat and those of electricity.

Although the substitution method has a clear link with consequential analyses, it has been often applied in the literature for attributional LCAs with goals similar to the one of this study [53–55]. By the substitution method, the impact of the main product is obtained by subtracting the impact of the marginally avoided secondary products from the impacts of the overall system [21,56]. In particular, the main product is defined as the one providing the highest share of revenues within the analyzed product system (physical/economic significance) [21].

A sensitivity analysis was performed to understand the influence of the method on the results of the study. This analysis explored the variation of the results when applying economic allocation and the substitution method for the WC scenario, ORC scenario, and separate productions of heat and electricity.

When applying substitution, the first step is identifying the main product. Based on the economic heat/electricity ratio (see Table 2), electricity is the main product for the HBP. The production of heat by the HBP technology can marginally avoid the production of heat from natural gas on the market (Heat, central or small-scale, natural gas {CH}|heat production, natural gas, at boiler condensing modulating < 100 kW|Conseq from ecoinvent 3.4).

On the other hand, the HBP heat could also avoid the production of heat by an average biomass boiler (Heat, district, or industrial, other than natural gas {CH}|heat production, softwood chips from the forest, at furnace 300 kW|APOS from ecoinvent 3.4). The choice of a biomass boiler as substituted technology can be considered as an "alternative activity allocation" i.e., a form of "proxy-based disaggregation" [21]. This type of allocation is performed through the subtraction of impacts but differs from the substitution performed in consequential LCAs because it is not based on modeling of marginality [21]. Instead, this allocation takes as substituted processes the ones providing "primary productions of identical products and not of products that fall under different categories" [21]. This approach might, therefore, be an option also in attributional LCAs when reflecting the underlying physical relationship between the main and subsidiary products [21]. This sensitivity analysis considered both approaches, the substitution of a marginal activity (heat from a natural gas condensing boiler) and the substitution of an alternative activity (heat from a biomass boiler, marked as (a) in Figures 9 and 10).

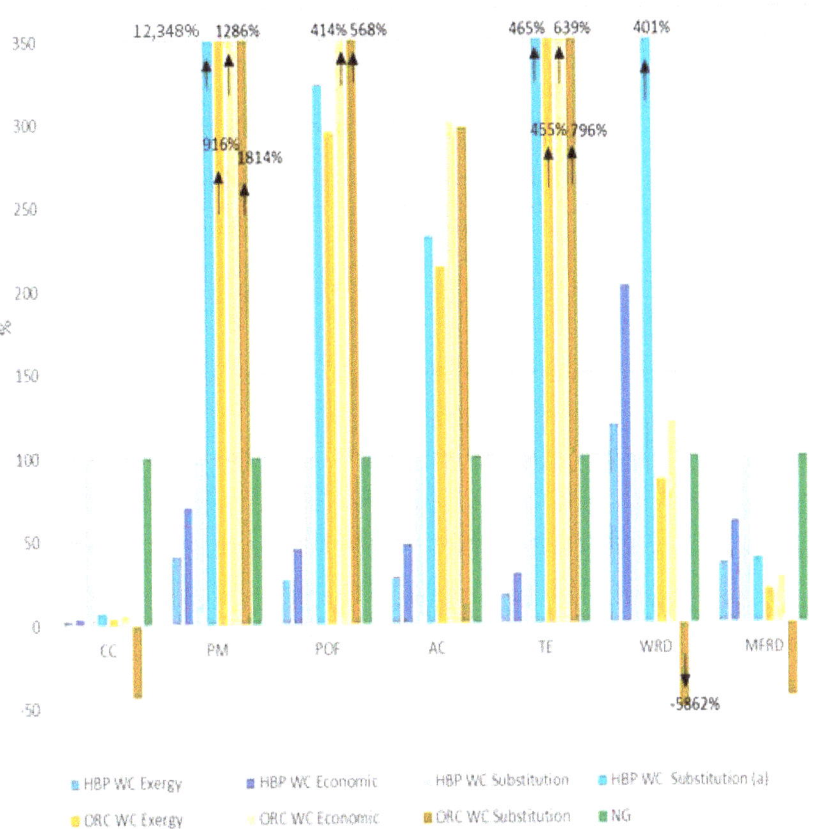

Figure 9. Sensitivity on allocation method for the generation of 1 MJ with HBP technology and competing technologies. Boiler running with natural gas taken as 100%. ORC = Organic Rankine Cycle, NG = Natural gas boiler, (a) = substitution of heat from a biomass boiler.

Based on the economic heat/electricity ratio of the ORC technology (see Table 2), heat is the main product for the ORC technology. In the case of the ORC, the electricity produced from the ORC avoids the production of marginal electricity from the electricity mix (this process is represented in the model by the ecoinvent dataset Electricity, high voltage {DE}|market for|Conseq).

The sensitivity analysis (see Figures 9 and 10) indicated that, compared to exergy allocation, the economic allocation method apportions more impacts on heat (+70% in every category) while it decreases by 23% the impacts of electricity. The same applies to the ORC technology (+40% and −36% respectively for heat and electricity). For CHPs, it is therefore important to show the impacts for both heat and electricity when an allocation method is applied, so that a full picture of its environmental impacts is provided.

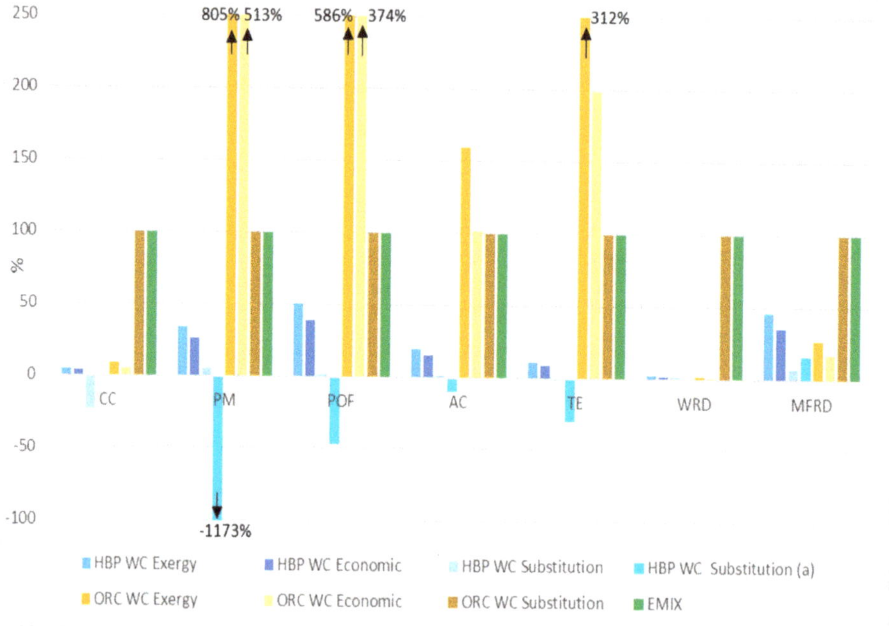

Figure 10. Sensitivity on allocation method for the generation of 1 kWh with HBP technology and competing technologies (1 kWh electricity). Electricity mix taken as 100%. ORC = Organic Rankine Cycle, EMIX = Electricity mix, (a) = substitution of heat from a biomass boiler.

On the other hand, the conclusions of the comparative assessment did not change when applying exergy or economic allocation methods. This was true for all three comparisons: (1) between HBP with wood chips and ORC with wood chips, (2) between the three different biomass fuels scenarios and (3) between the HBP and the separate productions. For instance, the impact of the HBP per MJ of heat with both allocation methods was lower than for ORC in climate change, particulate matter, photochemical ozone formation, acidification, and terrestrial eutrophication, but it was higher in the two categories concerning the depletion of resources (see Figure 9). On the other hand, the percentages of potential environmental impact savings or intensifications compared to separate production can change significantly. For example, for climate change, the savings of impact of the HBP compared to ORC was 42% when using exergy allocation while it was decreased to 30% with economic allocation. However, for particulate matter, there was no difference.

Concerning substitution (see Figures 9 and 10), the variations compared to other allocation methods were small or large depending on the impact category considered and the type of substitution applied. Moreover, both types of substitution approaches and the alternative activity method led to negative results in some impact categories. This last aspect highlights that the modeling was not consistent with the attributional goal of the study, which is not aimed at assessing a change in demand, and therefore, it should provide negative emissions for a single product of a multifunctional process whose overall impact is positive [21]. When a physically/economically significant product (the substituted function was 42% of total revenues for HBP and 34% for ORC) is substituted in attributional LCAs (by assuming that its impact corresponds to the one that would be replaced in the market), the results are often not aligned with other allocation methods and contrasts with the attributional aim of the LCA. This aspect emerges clearly when multiple impact categories are investigated in the same LCA study resulting in conclusions in contrast with other allocation methods and of difficult interpretation.

3.4. Sensitivity Analysis on Potentially Sensitive Parameters

3.4.1. Internal Parameters

The results of the analysis indicated that the production of the biomass fuel (23–78% for the baseline scenario WC, depending on the category) and the SOFC stacks (10–43%) have a high contribution to the total impacts.

The ecoinvent dataset used for wood chips included both wood chips obtained as by-products of sawmill activities (15%) and from forest management (85%). To reduce the environmental impact, a scenario with only sawmill wood chips as fuel could be used. This type of wood chips presents a lower impact compared to wood chips from forest management because an important percentage of the impact of the upstream activities occurring in the forests is allocated to the main products of the sawmills. This scenario was assessed by sensitivity analysis to estimate the potential variation in the impact of the HBP (see the second column of Table 5). By using only sawmill wood chips, the environmental impact of HBP technology can be significantly reduced (indicatively by 10–40%).

Table 5. Sensitivity analysis on the reduction of the environmental impact of the HBP technology by either increasing the SOFC stack lifetime or using only wood chips produced as industrial by-products.

Impact Category	Industrial Wood Chips Only (Measured in % Variation)	Longer SOFC Stack Lifetime (Measured in % Variation)
Climate change (CC)	−37%	−4%
Particulate matter (PM)	−15%	−6%
Photochemical ozone formation (POF)	−47%	−2%
Acidification (AC)	−16%	−7%
Terrestrial eutrophication (TE)	−37%	−4%
Water resource depletion (WRD)	+8%	−10%
Mineral, fossil, and renewable resource depletion (MFRD)	−10%	−3%

Alternatively, the impact of HBP technology could be improved by acting on the SOFC stack. Since the stack needs to be replaced every five years, the environmental impact could be improved by increasing the SOFC stack lifetime and therefore reducing the number of replacements over the plant lifetime. The second column of Table 5 shows the reduction of the environmental impact that could potentially be achieved by increasing the lifetime of the SOFC from five to seven years. This would lead to a decrease between 2% and 10% of the impacts of the wood chips scenario (baseline).

3.4.2. External Parameters

Since the technology will be deployed after 2025, it is important to explore how the comparative evaluation will change taking into account the current trends of decarbonization, which should lead to a decrease in the share of coal-produced electricity by shifting to renewables. In particular, the expected decarbonization of the European electricity grid will diminish the environmental benefits of HBP technology.

To assess this variation, the electricity mix based on two future scenarios for 2030 were considered: the EU reference scenario for Germany [57] and the IEA current policy scenario [58]. Due to the unavailability of the IEA current policy scenario for Germany, the IEA average mix of 2030 for the EU was taken as a proxy. This second scenario represents a more decarbonized electricity sector and includes other countries where the HBP could be commercialized. In particular, the IEA current policy scenario has only 13.7% coal and 44.7% renewables. The future savings of environmental impact allowed by the HBP is shown in the two columns on the right in Table 6. The environmental savings from the HBP technology will be only slightly affected (order of 5% overall) by the change expected in the electricity mix for 2030.

Table 6. Sensitivity analysis on the savings of environmental impacts of 1 kWh of electricity produced by the HBP compared to the grid electricity mix (EMIX).

Impact Category	EMIX Germany, Ecoinvent 3.4		Future EMIX Germany, EU Reference Scenario 2030		Future EMIX EU, IEA Current Policy Scenario 2030	
	1 kWh Electricity	Savings (%) HBP vs. EMIX	1 kWh Electricity	Savings (%) HBP vs. EMIX	1 kWh Electricity	Savings (%) HBP vs. EMIX
Climate change (kg CO_2eq)	6.41×10^{-1}	−95%	5.61×10^{-1}	−95%	3.96×10^{-1}	−92%
Particulate matter (kg PM2.5 eq)	7.87×10^{-5}	−66%	7.84×10^{-5}	−66%	6.74×10^{-5}	−60%
Photochemical ozone formation (kg NMVOC eq)	6.02×10^{-4}	−49%	6.06×10^{-4}	−50%	4.98×10^{-4}	−39%
Acidification (molc H+ eq)	1.58×10^{-3}	−80%	1.35×10^{-3}	−77%	1.12×10^{-3}	−72%
Terrestrial eutrophication (molc N eq)	4.33×10^{-3}	−89%	3.86×10^{-3}	−87%	3.49×10^{-3}	−86%
Water resource depletion (m^3 water eq)	2.75×10^{-3}	−98%	1.32×10^{-4}	−54%	4.47×10^{-4}	−87%
Mineral, fossil and renewable resource depletion (kg Sb eq)	7.47×10^{-6}	−54%	7.89×10^{-6}	−56%	7.41×10^{-6}	−54%

4. Conclusions

This article presented the first life cycle assessment (LCA) of a novel technology integrating biomass gasification and SOFC technologies. This technology is currently under development in the H2020 HiEff-BioPower (HBP) project and allows for the use of various biomass types as feedstock. This LCA assessed the environmental impacts when operating the technology with three different fuels: wood chips, wood pellets, and *Miscanthus* pellets. The impact of producing heat and electricity with the HBP technology was compared to the state of the art competing technologies. The results showed that most of the impacts of producing heat and electricity with the HBP technology are generated during the production (including transportation) of the biomass fuels (between 23% and 99% of the total impacts depending on the category and the fuel). The use of wood chips as fuel generates much lower impacts per functional unit than the operation with wood pellets (11–70% lower) and *Miscanthus* pellets (9–99% lower), in all impact categories. The next highest contributor to the life cycle environmental impacts is the SOFC stack, due to both the high energy intensity (especially in electricity consumption) and material intensities of its manufacturing processes, and its short lifetime (the stack should be replaced every 5 years). Beyond increasing the fuel efficiency of the technology and therefore reducing the consumption of biomass fuels, the main recommendation to technology developers would be to increase the lifetime of the SOFC stack. Increasing the SOFC stack lifetime could decrease the environmental impacts of 2–10%, depending on the category.

The comparison of the HBP technology with separate productions of heat and electricity (from natural gas condensing boilers and the German electricity grid) indicated significantly lower impacts for the HBP technology, especially in climate change (86%/94% lower), photochemical ozone formation (−43%/−70%), acidification (−37%/−56%) and terrestrial eutrophication (−43%/−63%). Overall, HBP showed also better performance than ORCs, as they have higher exergy efficiencies and almost zero particulate emissions resulting in 86–96% lower impact in the category particulate matter.

The sensitivity analysis on the allocation method for heat and electricity provided useful insights for the choice of allocation methods in CHP plants, and led to the following recommendations: (1) the attributional LCAs of CHPs should always provide the results for both heat and electricity to allow for better interpretation of results, independently of the allocation method, (2) LCA results from different CHP plants should not be compared if they assumed different allocation approaches and (3) substitution is not recommended in attributional LCA (especially if the substituted product is not a minor by-product) because it provides results which are not in line with the attributional aim (e.g., negative emissions) and lead to conclusions in contracts with the ones from applying allocation methods which are proven to be a good proxy of physical causality for CHPs and therefore preferable.

Author Contributions: Conceptualization, C.M. and B.C.; data curation, C.M., V.R., T.G., T.B. and I.O.; formal analysis, C.M., B.C. and V.R.; funding acquisition, M.J. and L.S.; investigation, C.M., B.C., V.R. and L.S.; Methodology, C.M., B.C., M.J. and L.S.; project administration, B.C., M.J., T.B., I.O., and L.S.; resources, B.C.; software, C.M. and V.R.; supervision, B.C., M.J. and L.S.; validation, B.C., T.G., M.J., T.B., I.O. and L.S.; visualization, B.C., M.J., and L.S.; writing—original draft, C.M. and V.R.; writing—review and editing, B.C., T.G., M.J., T.B., I.O. and L.S. All authors have read and agreed to the published version of the manuscript.

Funding: This LCA study has been developed within the HiEff-BioPower project, which received funding by the European Union Horizon2020 Programme under Grant agreement number 727330.

Acknowledgments: The authors express their gratitude to the other project partners and all the members of the consortium.

Conflicts of Interest: The authors declare no conflict of interest.

Appendix A

Table A1 presents the characterized environmental impacts per functional unit of heat and electricity for each fuel scenario.

Table A1. Cradle-to-grave environmental impacts per functional unit and biomass fuel. WC = wood chips, WP = wood pellets, MP = *Miscanthus* pellets.

Impact Category	1 MJ of Heat			1 kWh of Electricity		
	Wood Chips	Wood Pellets	*Miscanthus* Pellets	Wood Chips	Wood Pellets	*Miscanthus* Pellets
Climate change (kg CO_2eq)	1.52×10^{-3}	3.42×10^{-3}	4.36×10^{-3}	3.05×10^{-2}	6.87×10^{-2}	8.77×10^{-2}
Particulate matter (kg PM2.5 eq)	1.34×10^{-6}	4.19×10^{-6}	3.07×10^{-6}	2.70×10^{-5}	8.41×10^{-5}	6.17×10^{-5}
Photochemical ozone formation (kg NMVOC eq)	1.51×10^{-5}	1.71×10^{-5}	1.66×10^{-5}	3.04×10^{-4}	3.43×10^{-4}	3.33×10^{-4}
Acidification (molc H+ eq)	1.57×10^{-5}	2.87×10^{-5}	3.49×10^{-5}	3.15×10^{-4}	5.78×10^{-4}	7.02×10^{-4}
Terrestrial eutrophication (molc N eq)	2.46×10^{-5}	5.72×10^{-5}	7.96×10^{-5}	4.93×10^{-4}	1.15×10^{-3}	1.60×10^{-3}
Water resource depletion (co^2 water eq)	3.01×10^{-6}	1.02×10^{-5}	3.36×10^{-4}	6.03×10^{-5}	2.04×10^{-4}	6.76×10^{-3}
Mineral, fossil and renewable resource depletion (kg Sb eq)	1.71×10^{-7}	2.80×10^{-7}	3.71×10^{-7}	3.43×10^{-6}	5.63×10^{-6}	7.46×10^{-6}

References

1. Havukainen, J.; Nguyen, M.T.; Väisänen, S.; Horttanainen, M. Life cycle assessment of small-scale combined heat and power plant: Environmental impacts of different forest biofuels and replacing district heat produced from natural gas. *J. Clean. Prod.* **2018**, *172*, 837–846. [CrossRef]

2. Boschiero, M.; Cherubini, F.; Nati, C.; Zerbe, S. Life cycle assessment of bioenergy production from orchards woody residues in Northern Italy. *J. Clean. Prod.* **2016**, *112*, 2569–2580. [CrossRef]

3. Bloess, A.; Schill, W.P.; Zerrahn, A. Power-to-heat for renewable energy integration: A review of technologies, modeling approaches, and flexibility potentials. *Appl. Energy* **2018**, *212*, 1611–1626. [CrossRef]

4. Lombardi, F.; Rocco, M.V.; Colombo, E. A multi-layer energy modelling methodology to assess the impact of heat-electricity integration strategies: The case of the residential cooking sector in Italy. *Energy* **2019**, *170*, 1249–1260. [CrossRef]

5. Chiaroni, D.; Chiesa, M.; Chiesa, V.; Franzò, S.; Frattini, F.; Toletti, G. Introducing a new perspective for the economic evaluation of industrial energy efficiency technologies: An empirical analysis in Italy. *Sustain. Energy Technol. Assess.* **2016**, *15*, 1–10. [CrossRef]

6.   Paletto, A.; Bernardi, S.; Pieratti, E.; Teston, F.; Romagnoli, M. Assessment of environmental impact of biomass power plants to increase the social acceptance of renewable energy technologies. *Heliyon* **2019**, *5*, e02070. [CrossRef]

7.   González-García, S.; Bacenetti, J. Exploring the production of bio-energy from wood biomass. Italian case study. *Sci. Total Environ.* **2019**, *647*, 158–168. [CrossRef] [PubMed]

8.   Bacenetti, J.; Fusi, A.; Azapagic, A. Environmental sustainability of integrating the organic Rankin cycle with anaerobic digestion and combined heat and power generation. *Sci. Total Environ.* **2019**, *658*, 684–696. [CrossRef] [PubMed]

9.   Götz, T.; Saurat, M.; Kaselofsky, J.; Obernberger, I.; Brunner, T.; Weiss, G.; Bellostas, B.C.; Moretti, C. First Stage Environmental Impact Assessment of a New Highly Efficient and Fuel Flexible Medium-scale CHP Technology Based on Fixed-bed Updraft Biomass Gasification and a SOFC. In Proceedings of the 27th European Biomass Conference and Exhibition, Lisbon, Portugal, 27–30 May 2019; pp. 1586–1594.

10.   Brunner, T.; Biedermann, F.; Obernberger, I.; Hirscher, S.; Schöch, M.; Milito, C.; Leibold, H.; Sitzmann, J.; Megel, S.; Hauth, M.; et al. Development of a new highly efficient and fuel flexible medium-scale CHP technology based on fixed-bed updraft biomass gasification and a SOFC. In Proceedings of the 26th European Biomass Conference and Exhibition: EUBCE 2018, Copenhagen, Denmark, 14–17 May 2018; Volume 72, pp. 249–267.

11.   European Commission. Technology Readiness Levels (TRL). Horizon 2020–Work Programme 2014–2015 General Annexes, Extract from Part 19-Commission Decision C. Available online: https://ec.europa.eu/research/participants/data/ref/h2020/wp/2014_2015/annexes/h2020-wp1415-annex-g-trl_en.pdf (accessed on 20 May 2020).

12.   Chianese, S.; Fail, S.; Binder, M.; Rauch, R.; Hofbauer, H.; Molino, A.; Blasi, A.; Musmarra, D. Experimental investigations of hydrogen production from CO catalytic conversion of tar rich syngas by biomass gasification. *Catal. Today* **2016**, *277*, 182–191. [CrossRef]

13.   Fail, S.; Diaz, N.; Benedikt, F.; Kraussler, M.; Hinteregger, J.; Bosch, K.; Hackel, M.; Rauch, R.; Hofbauer, H. Wood gas processing to generate pure hydrogen suitable for PEM fuel cells. *ACS Sustain. Chem. Eng.* **2014**, *2*, 2690–2698. [CrossRef]

14.   Biedermann, F.; Brunner, T.; Obernberger, I.; Weiß, G. *D8. 3: Preliminary Techno-Economic Performance Analysis of the New Technologies*; H2020 Hieff-BioPower project deliverable; BIOS: Graz, Austria, October 2017.

15.   Evangelisti, S.; Lettieri, P.; Clift, R.; Borello, D. Distributed generation by energy from waste technology: A life cycle perspective. *Process Saf. Environ. Prot.* **2015**, *93*, 161–172. [CrossRef]

16.   Lee, Y.D.; Ahn, K.Y.; Morosuk, T.; Tsatsaronis, G. Environmental impact assessment of a solid-oxide fuel-cell-based combined-heat-and-power-generation system. *Energy* **2015**, *79*, 455–466. [CrossRef]

17.   Rillo, E.; Gandiglio, M.; Lanzini, A.; Bobba, S.; Santarelli, M.; Blengini, G. Life Cycle Assessment (LCA) of biogas-fed Solid Oxide Fuel Cell (SOFC) plant. *Energy* **2017**, *126*, 585–602. [CrossRef]

18.   ISO. ISO 14044, Environmental management—Life cycle assessment—Requirements and guidelines. International Standard Organization. *Environ. Manag.* **2006**, *4*, 307.

19.   Adams, P.W.R.; Manus, M.C. Small-scale biomass gasification CHP utilisation in industry: Energy and environmental evaluation. *Sustain. Energy Technol. Assess.* **2014**, *6*, 129–140. [CrossRef]

20.   Mehmeti, A.; McPhail, S.J.; Pumiglia, D.; Carlini, M. Life cycle sustainability of solid oxide fuel cells: From methodological aspects to system implications. *J. Power Sources* **2016**, *325*, 772–785. [CrossRef]

21.   Majeau-Bettez, G.; Dandres, T.; Pauliuk, S.; Wood, R.; Hertwich, E.; Samson, R.; Strømman, A.H. Choice of allocations and constructs for attributional or consequential life cycle assessment and input-output analysis. *J. Ind. Ecol.* **2018**, *22*, 656–670. [CrossRef]

22.   Vera, I.; Hoefnagels, R.; van der Kooij, A.; Moretti, C.; Junginger, M. A carbon footprint assessment of multi-output biorefineries with international biomass supply: A case study for the Netherlands. *Biofuels Bioprod. Biorefining* **2020**, *14*, 198–224. [CrossRef]

23.   ISO 14040. *Environmental Management—Life Cycle Assessment—Principles and Framework*; Technical Committee ISO/TC 207; ISO: Geneva, Switzerland, 2006.

24.   Pelletier, N.; Ardente, F.; Brandão, M.; De Camillis, C.; Pennington, D. Rationales for and limitations of preferred solutions for multi-functionality problems in LCA: Is increased consistency possible? *Int. J. Life Cycle Assess.* **2015**, *20*, 74–86. [CrossRef]

25. Corona, B.; Shen, L.; Junginger, M. *Preliminary Market Study for Europe: Detailed Market Assessment of 4 EU Member State Markets*; H2020 Hieff-BioPower project deliverable; Utrecht University: Utrecht, The Netherlands, 2018.

26. ILCD. ILCD Handbook—General guide on LCA—Detailed guidance. *Constraints* **2010**, *15*, 524–525.

27. Höök, M.; Tang, X. Depletion of fossil fuels and anthropogenic climate change—A review. *Energy Policy* **2013**, *52*, 797–809. [CrossRef]

28. Conibear, L.; Butt, E.W.; Knote, C.; Arnold, S.R.; Spracklen, D.V. Residential energy use emissions dominate health impacts from exposure to ambient particulate matter in India. *Nat. Commun.* **2018**, *9*, 1–9. [CrossRef]

29. Broeren, M.L.M.; Zijp, M.C.; Waaijers-van der Loop, S.L.; Heugens, E.H.W.; Posthuma, L.; Worrell, E.; Shen, L. Environmental assessment of bio-based chemicals in early-stage development: A review of methods and indicators. *Biofuels Bioprod. Biorefining* **2017**, *11*, 701–718. [CrossRef]

30. Hartmann, D.L.; Tank, M.G.K.; Rusticucci, M. *IPCC Fifth Assessment Report, Climatie Change 2013: The Physical Science Basis*; IPCC AR5; IPCC: Geneva, Switzerland, 2013.

31. Rabl, A.; Spadaro, J.V.; Holland, M. Description of the RiskPoll software. In *How Much Is Clean Air Worth*; Cambridge University Press: Cambridge, UK, 2014.

32. Van Zelm, R.; Huijbregts, M.A.J.; den Hollander, H.A.; van Jaarsveld, H.A.; Sauter, F.J.; Struijs, J.; van Wijnen, H.J.; van de Meent, D. European characterization factors for human health damage of PM10 and ozone in life cycle impact assessment. *Atmos. Environ.* **2008**, *42*, 441–453. [CrossRef]

33. Posch, M.; Seppälä, J.; Hettelingh, J.P.; Johansson, M.; Margni, M.; Jolliet, O. The role of atmospheric dispersion models and ecosystem sensitivity in the determination of characterisation factors for acidifying and eutrophying emissions in LCIA. *Int. J. Life Cycle Assess.* **2008**, *13*, 477. [CrossRef]

34. Frischknecht, R.; Steiner, R.; Arthur, B.; Norbert, E.; Gabi, H. Swiss Ecological Scarcity Method: The New Version 2006. Available online: https://www.researchgate.net/publication/237790160_Swiss_Ecological_Scarcity_Method_The_New_Version_2006 (accessed on 28 May 2020).

35. Van Oers, L.; de Koning, A.; Guinée, J.B.; Huppes, G. Abiotic Resource Depletion in LCA. 2002. Available online: https://www.leidenuniv.nl/cml/ssp/projects/lca2/report_abiotic_depletion_web.pdf (accessed on 20 May 2020).

36. Mackenzie, S.G.; Leinonen, I.; Kyriazakis, I. The need for co-product allocation in the life cycle assessment of agricultural systems—Is "biophysical" allocation progress? *Int. J. Life Cycle Assess.* **2017**, *22*, 128–137. [CrossRef]

37. Azapagic, A.; Clift, R. Allocation of Environmental Burdens in Co-product Systems: Product-related Burdens (Part 1). *Int. J. Life Cycle Assess.* **1999**, *4*, 357–369. [CrossRef]

38. Moretti, C.; Junginger, M.; Shen, L. Environmental life cycle assessment of polypropylene made from used cooking oil. *Resour. Conserv. Recycl.* **2020**, *157*, 104750. [CrossRef]

39. Primas, A. *Life Cycle Inventories of New CHP Systems*; Ecoinvent: Zurich, Switzerland, 2007.

40. European Commission. Proposal for a Directive of the European Parliament and of the Council on the promotion of the use of energy from renewable sources. *Off. J. Eur. Union* **2016**, *5*, 2009.

41. EUROSTAT Energy Data. Available online: https://ec.europa.eu/eurostat/web/energy/data (accessed on 15 November 2019).

42. Perić, M.; Komatina, M.; Antonijević, D.; Bugarski, B.; Dželetović, Ž. Life Cycle Impact Assessment of Miscanthus Crop for Sustainable Household Heating in Serbia. *Forests* **2018**, *9*, 654. [CrossRef]

43. Caslin, B.; Finnan, J.; Easson, L. Miscanthus Best Practice Guidelines; Agriculture and Food Development Authority. Available online: https://www.teagasc.ie/publications/2011/315/MiscanthusBestPractice.pdf (accessed on 13 April 2019).

44. Staffell, I.; Ingram, A.; Kendall, K. Energy and carbon payback times for solid oxide fuel cell based domestic CHP. *Int. J. Hydrog. Energy* **2012**, *37*, 2509–2523. [CrossRef]

45. BWF Tec GmbH & Co. KG. *Pyrotex ®KE. More Than Just Hot Gas De-Dusting*; BWF Envirotec: Offingen, Germany, 2017.

46. Biganzoli, L.; Rigamonti, L.; Grosso, M. LCA evaluation of packaging re-use: The steel drums case study. *J. Mater. Cycles Waste Manag.* **2019**, *21*, 67–78. [CrossRef]

47. Rigamonti, L.; Grosso, M.; Sunseri, M.C. Influence of assumptions about selection and recycling efficiencies on the LCA of integrated waste management systems. *Int. J. Life Cycle Assess.* **2009**, *14*, 411–419. [CrossRef]

48. Valente, A.; Iribarren, D.; Dufour, J. End of life of fuel cells and hydrogen products: From technologies to strategies. *Int. J. Hydrog. Energy* **2019**, *44*, 20965–20977. [CrossRef]

49. Milà i Canals, L.; Bauer, C.; Depestele, J.; Dubreuil, A.; Freiermuth Knuchel, R.; Gaillard, G.; Michelsen, O.; Müller-Wenk, R.; Rydgren, B. Key Elements in a Framework for Land Use Impact Assessment Within LCA (11 pp). *Int. J. Life Cycle Assess.* **2007**, *12*, 5–15. [CrossRef]

50. Strazza, C.; Del Borghi, A.; Costamagna, P.; Gallo, M.; Brignole, E.; Girdinio, P. Life Cycle Assessment and Life Cycle Costing of a SOFC system for distributed power generation. *Energy Convers. Manag.* **2015**, *100*, 64–77. [CrossRef]

51. Sadhukhan, J. Distributed and micro-generation from biogas and agricultural application of sewage sludge: Comparative environmental performance analysis using life cycle approaches. *Appl. Energy* **2014**, *122*, 196–206. [CrossRef]

52. Osman, A.; Ries, R. Life cycle assessment of electrical and thermal energy systems for commercial buildings. *Int. J. Life Cycle Assess.* **2007**, *12*, 308–316. [CrossRef]

53. Schrijvers, D.L.; Loubet, P.; Sonnemann, G. Developing a systematic framework for consistent allocation in LCA. *Int. J. Life Cycle Assess.* **2016**, *21*, 976–993. [CrossRef]

54. Sandin, G.; Røyne, F.; Berlin, J.; Peters, G.M.; Svanström, M. Allocation in LCAs of biorefinery products: Implications for results and decision-making. *J. Clean. Prod.* **2015**, *93*, 213–221. [CrossRef]

55. Anne Renouf, M.; Pagan, R.J.; Wegener, M.K. Life cycle assessment of Australian sugarcane products with a focus on cane processing. *Int. J. Life Cycle Assess.* **2011**, *16*, 125–137. [CrossRef]

56. Cherubini, F.; Strømman, A.H.; Ulgiati, S. Influence of allocation methods on the environmental performance of biorefinery products—A case study. *Resour. Conserv. Recycl.* **2011**, *55*, 1070–1077. [CrossRef]

57. Carpos, P.; De Vita, A.; Tasios, N.; Siskos, P.; Kannavou, M.; Preopoulos, A.; Evangelopoulou, S.; Zampara, M.; Papadopoulos, D.; Nakos, C.; et al. *EU Reference Scenario 2016—Energy, Transport and GHG Emissions—Trends to 2050*; European Commission: Brussels, Belgium, 2016; ISBN 978-92-79-52373-1.

58. IEA. *World Energy Outlook 2018: Electricity*; IEA: Paris, France, 2018.

Article

Complete Data Inventory of a Geothermal Power Plant for Robust Cradle-to-Grave Life Cycle Assessment Results

Lorenzo Tosti [1,2], Nicola Ferrara [1,2], Riccardo Basosi [1,2,3],* and Maria Laura Parisi [1,2,3],*

[1]   R2ES Lab, Department of Biotechnology, Chemistry and Pharmacy, University of Siena, 53100 Siena, Italy; tosti@csgi.unifi.it (L.T.); ferrara33@student.unisi.it (N.F.)

[2]   CSGI—Consortium for Colloid and Surface Science, via della Lastruccia 3, 50019 Sesto Fiorentino, Italy

[3]   CNR-ICCOM, National Research Council—Institute for the Chemistry of OrganoMetallic Compounds, 50019 Sesto Fiorentino, Italy

*   Correspondence: riccardo.basosi@unisi.it (R.B.); marialaura.parisi@unisi.it (M.L.P.)

Received: 14 May 2020; Accepted: 29 May 2020; Published: 3 June 2020

Abstract: Technologies to produce electric energy from renewable geothermal source are gaining increasing attention, due to their ability to provide a stable output suitable for baseload production. Performing life cycle assessment (LCA) of geothermal systems has become essential to evaluate their environmental performance. However, so far, no documented nor reliable information has been made available for developing robust LCA studies. This work provides a comprehensive inventory of the Italian Bagnore geothermal power plants system. The inventory is based exclusively on primary data, accounting for every life cycle stage of the system. Data quality was assessed by means of a pedigree matrix. The calculated LCA results showed, with an overall low level of uncertainty (2–3%), that the commissioning and operational phases accounted for more than 95% of the environmental profile. Direct emissions to atmosphere were shown to be the major environmental impact, particularly those released during the operational phase (84%). The environmental performances comparison with the average Italian electricity mix showed that the balance is always in favor of geothermal energy production, except in the climate change impact category. The overall outcome confirms the importance, for flash technology employing fluid with a high concentration of gas content, of using good quality primary data to obtain robust results.

Keywords: geothermal energy; flash technology; Bagnore power plant; life cycle assessment; pedigree matrix

---

1. Introduction

The European Commission (EC) is promoting the transition of the European Union (EU) to a highly energy efficient and low-carbon economy system [1]. Energy production from renewable energy sources (RES), saving energy and natural resources, as well as reducing carbon dioxide (CO_2) emissions while managing wastes, are pivotal actions to enable such a transition [2]. The EC adopted the "2030 climate and energy framework" in 2014, which has been subsequently revised in 2018 to include broader targets and policy objectives on greenhouse gasses (GHG) emission reduction for the period from 2021 to 2030. The targets for RES and energy efficiency are set to at least 40% cuts in greenhouse gas emissions (from 1990 levels), at least a 32% share for renewable energy, and at least a 32.5% improvement in energy efficiency [3]. On November 2018, the EC presented the analytical foundation for the development of an EU Long Term Strategy for climate and energy policy and a political vision for achieving a Net Zero economy by 2050 [4]. In this context, power generation has been identified as one of the sectors with the highest potential to decarbonize. To ensure that the EU targets are met, EU legislation [5] requires that each Member State drafts a 10-year National Energy and Climate Plan (NECP), setting out how to reach its national targets. The Italian NECP [6], largely built on the 2017 Italian Energy Strategy, broadly meets the requirements set by the Regulation. The draft of the plan has been positively judged by EC, as it includes an extensive list of 101 policies and measures. These would be enough for Italy to meet the above targets, with a particularly important contribution coming from the objective of gradually phasing out coal for electricity generation by 2025. The draft plan qualitatively mentions the interactions with air quality and air emissions policy, specifically in the context of the proposed contribution expressed as 30% share of energy from RES in gross final consumption of energy in 2030. Electric energy production from RES, particularly those not emitting into the atmosphere during the operational phase like solar, wind and hydro, will play a key role in achieving such an ambitious objective. Biomass and geothermal can also play a role in replacing fossils toward a more sustainable development, but they are not exempt from drawbacks concerning CO_2 emissions [7]. As geothermal energy has big potential for development [8], it is becoming important to explore the state of the art of the technology in terms of a benefit/cost ratio from an environmental point of view. Among RES, geothermal energy is considered a competitive energy source, because of its independence from seasonal and climatic conditions [9], ensuring reliable performances peculiar to non-renewable sources. Geothermal power plants can provide a stable production output, unaffected by the external environment, resulting in high capacity factors (ranging from 60% to 90%), and making the technology suitable for baseload production [10]. The technologies for power production from geothermal resource exploitation depend on the quality of the geothermal field, which, in general, increases with its enthalpy, typically spanning from liquid-only to steam-only (i.e., dry steam) reservoirs. Naturally occurring geothermal systems, known as hydro-thermal, are characterized by a resource fluid condition that can be considered directly available. By contrast, enhanced geothermal systems (EGS) aim to produce hot water at locations where natural aquifers are not present by developing an "engineered reservoir". This technology has received significant attention, because it allows the exploitation of geothermal energy virtually anywhere. Hydrothermal (mono, double or triple flash and dry steam) plants account for around 85% of the global geothermal power generation. In 2018, this was an estimated 90 TWh, while the cumulative capacity reached 14 GW [11]. Around 14% of the global electricity production is due to a different technology based on binary cycles [12]. This technology often exploits the total re-injection of non-condensable gases (NCGs) with some environmental advantages, despite a significant decrease in efficiency and larger land occupancy [8]. In this context, the concern about the environmental performance of geothermal energy exploitation has been growing in recent years, due to the expected increase of power production from geothermal sources [13].

Life Cycle Assessment (LCA) methodology is one of the most reliable and powerful tools to assess the environmental performance of power generation systems, capable of providing results that cover several environmental aspects, thus approaching the system in a more comprehensive and holistic way [14–16].

Even though LCA has been applied for quite a long time now to energy-producing systems, the field of geothermal energy exploitation still lacks primary data. Only a few studies have been aimed at determining the environmental profile of currently operating geo-thermoelectric installations in Italy [17–19] and in Iceland [20,21].

The relative complexity and high dependency on geomorphological factors of the geothermal energy source also contributes to the scarcity of specific information. Reviews performed by several authors [22–24] underlined the inaccuracy due to the lack of primary data. This trend is even more evident in harmonization [25–27], which needs to deal with very large variability, making the elaboration of reliable eco-profile very difficult [28,29].

The consequence is that papers which analyze geothermal power plants mostly use secondary data, forcing the authors to rely on the general literature data, which are often not adequately representative of the technology and of the investigated system [30,31].

Recently, special attention has been paid to the evaluation of environmental performances of EGS [32,33]. However, at present, hydro-thermal systems dominate current electricity generation in the geothermal sector, and the exploitation of this type of reservoir is predicted to become dominant in the future [11,34]. This picture outlines the importance of assessing the life cycle environmental impacts of conventional geothermal technologies to make sustainable choices in the context of the electric energy production sector. To avoid uncertainties, a reliable and high-quality life cycle inventory of a flash installation is needed. The only current source of data is the one provided in the study by Karlsdottir et al. [35].

The scope of the present work is to provide a high-quality, complete and documented life cycle inventory of a flash power plant, and to perform the LCA of electricity production from geothermal source with a cradle to grave approach, and to evaluate how much uncertainty of data is reflected on the final LCA results. The quality of data was assessed employing a so-called pedigree or uncertainty matrix. The Italian Bagnore power plant was selected as one of the most representative flash-based conversion system power plants. This work has been made possible by the full availability of primary data which, according to our knowledge, is unique in the literature.

The Bagnore power plant system consists of three connected units, namely: Bagnore 3, the binary group of Bagnore 3 and Bagnore 4. To correctly assess the environmental footprint of these plants, it is necessary to consider them as a whole system, namely the Bagnore system. Bagnore power plants integrate two systems for atmospheric emissions abatement, namely the AMIS (i.e., the abatement system for mercury (Hg) and hydrogen sulfide (H_2S)) and the ammonia (NH_3) abatement system. The adoption of state-of-the-art management strategies by the operator, Enel Green Power (EGP), aims for the best trade-off between production performance and environmental compliance [36].

2. Materials and Methods

2.1. Power Plants Description

The production of electricity from geothermal exploitation in Italy is concentrated in the Tuscany region. Currently, all the geothermal power plants have been built and operated by EGP, which manages 37 productive units, allowing for the production of about 5.8 TWh/y.

The geothermal area in Tuscany is divided in four districts: Larderello, Lago and Radicondoli (halfway from the province of Siena, Grosseto and Pisa) and, in the south Tuscany, the area of Mount Amiata (between Grosseto and Siena) [37]. The area of Monte Amiata is composed by two productive geothermal fields, namely Bagnore and Piancastagnaio. The Bagnore field is characterized by the presence of 2 power plants: Bagnore 3 and Bagnore 4, entirely constructed and operated by EGP. Bagnore 3 is a flash plant with 20 MWe of installed power producing 170 GWh/y of electric energy. Additionally, the plant is powered by a 1 MWe Organic Rankine Cycle (ORC) unit, which provides 6.8 GWh/y of additional electric energy. Bagnore 4 is powered by two 20 MWe groups, which can input 367 GWh/y to the electric grid. Thus, the electricity production from the Bagnore field is about 544 GWh/y. In addition to electric generation, heat delivery is also achieved by exploiting residual heat after turbine expansion. The total heat delivered to the final users is about 32 GWh/y.

The two power plants are connected to each other to enhance the performance of the whole system. Such an enhancement is reached in both power production and environmental compatibility of the geothermal power plants [38]. A shared steam network powering the two power plants, allows the optimization of the available steam flow, thus, maximizing the power output. The shared steam network also improves the environmental footprint: in case of maintenance operations to one of the three productive units, it is possible to reroute the overflowing steam towards the operating units, thus avoiding free release into atmosphere. The operator also equipped the power plants with oversized AMIS system, able to treat 150% of the entering fluid for each turbine. Such oversizing allows the system to also abate the emissions during flow rerouting for maintenance operations.

2.2. LCA Methodological Approach

2.2.1. Goal and Functional Unit

The goal of the present LCA study is to assess the potential environmental impacts that are associated with the production of electricity from the geothermal power plants of Bagnore 3 and Bagnore 4. The functional unit selected is 1 kWh of net electricity produced. This study was conducted according to the requirements of ISO 14040 standard series [39,40] and the ILCD Handbook [41], following an attributional approach. The broader scope of the study was to provide insight and reference values on the environmental performances of an operating flash-based geothermal facility relying on a very detailed LCI. The intended application was to calculate the comprehensive eco-profile of the Bagnore power plants system. Data for building the life cycle inventory was obtained directly from EGP through accurate surveys and questionnaires.

2.2.2. System Boundaries

Figure 1 shows the system boundaries of the LCA study. The system modelling approach is cradle to grave, and it included the following phases: commissioning, operation and maintenance (which together constitute the use phase), decommissioning and end of life (EoL). The system boundaries were set up to the point where energy, in the form of electricity and heat, is produced from the plant. The distribution of energy was not considered. The life cycle phases included in the boundaries were modelled using foreground and background processes. The distinction between the foreground and background processes consists of the former being explicitly modelled for the investigated system employing data directly measured in situ, and is therefore highly representative of the technological and geographical situation of the studied system (primary data). Background processes are all the other processes for which data were retrieved from the Ecoinvent database version 3.5 [42] (secondary data). Background processes represent an average situation with a different level of geographical and technological representativeness, ranging from national to worldwide averages.

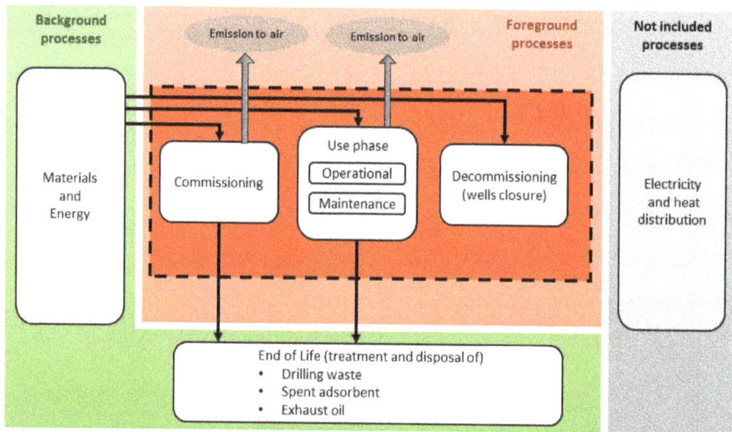

Figure 1. Graphical representation of the system boundaries considered in this study. A distinction is made between background processes that use secondary data (Ecoinvent database v3.5) and foreground processes that use primary data representative of the Bagnore power plant system. Electricity and heat distribution infrastructure and processes are not included in this study.

The geothermal power plant system was modelled in more detail, as follows:

(1) The commissioning phase included the drilling of production and injection wells and the well-pad construction; the construction of pipelines; the power plants building; all the equipment needed

for the power generation plant and the heating station, including the cooling towers, the ORC unit and the production of the working fluid used by the ORC system; the construction of AMIS was also included in the commissioning stage; on the contrary, the exploration and wells' testing stages were excluded from the analysis, due to lack of data.

(2) The operational phase included the production of sulfuric acid (H2SO4) used in the fluid treatment for the oxidation of (NH3); working fluid losses during normal operation of the ORC system were inventoried as direct emissions to air; the operational phase accounted for direct emissions of NCGs, Arsenic (As) and Hg to air. Maintenance activities included in the system were: AMIS maintenance, which involved the substitution of selenium based sorbent; the evaporative tower maintenance, which involved substitution of plastic parts (drift eliminator, fan); equipment maintenance, which includes lubricant oil refilling, substitution of metals components of various technical parts (i.e., turbine, compressors). More details are given in the Supplementary Materials in the "Inventory" sheet.

(3) When the geothermal power plant runs out of its lifetime, a decommissioning phase was assumed, which included exclusively the closing of the wells with cement. The activities of dismantling and recycling of machinery and equipment were excluded from the decommissioning stage, because they can be employed in other plants operated by EGP.

(4) Finally, the EoL phase included the treatment and disposal of drilling mud and of the spent sorbent from AMIS maintenance, as well as the treatment of exhaust oil from equipment maintenance activity.

2.2.3. LCA Key Modelling Parameters

This section reports the key modelling parameters of the geothermal plant (see Table 1), as well as the secondary data selection. The inventory of the Bagnore system is discussed in Section 3, where a general comparison in terms of data quality and coverage with the currently available LCI for Hellisheidi [35] is performed.

Table 1. Bagnore power plant system LCA key modelling parameters.

Geothermal Source Type		Hydrothermal
Energy generation technology		Flash
Final energy use		Electricity production
Average Reservoir Depth (shallow|deep) (m)		700|3000
Field Average Temperature (°C)		300–350
Parameter	Unit	Value
Installed power		
Electric	MWe	61
Thermal	MW$_{th}$	21.1
District Heating SUPPLY| RETURN temperature	°C	100|60
Net energy output (annual)		
Electric	GWh$_e$/y	544
Thermal	GWh$_{th}$/y	32
Predicted lifetime	Years	40
Total Energy Produced		
Electric	GWh$_e$	21,760
Thermal	GWh$_t$	1280
Production and injection wells	Number	8 production/6 injection
Total length drilled	Meters	31,823
Pipelines length	Meters	10,400
Load factor	%	99

Site specific data have always been used when possible, whereas background data were retrieved from the Ecoinvent database [42], with a preference for specific Italian dataset when available (i.e., for the electricity mix). When not available, average European or global dataset were selected.

2.2.4. Data Representativeness and Quality

The quality of collected data was assessed by means of the Ecoinvent data quality system [43]. Five indicators (i.e., reliability, completeness, temporal correlation, geographical correlation, further technological correlation) were assessed using a score from the best quality (score 1 corresponding to a verified measured data) to the worst (score 5 corresponding to not qualified/or estimate data). The complete description of indicators and scores is reported in Table 2.

Table 2. Data quality indicators and score description on scores [1].

Indicators	Scores				
	1	2	3	4	5
Reliability	Verified data based on measurements	Verified data partly based on assumptions or non-verified data based on measurements	Non-verified data partly based on qualified estimates	Qualified estimate (e.g., by industrial expert)	Non-qualified estimates
Completeness	Representative data from all sites relevant for the market considered, over and adequate period to even out normal fluctuations	Representative data from >50% of the sites relevant for the market considered, over an adequate period to even out normal fluctuations	Representative data from only some sites (<<50%) relevant for the market considered or >50% of sites but from shorter periods	Representative data from only one site relevant for the market considered or some sites but from shorter periods	Representativeness unknown or data from a small number of sites and from shorter periods
Temporal correlation	Less than 3 years of difference to the time period of the data set	Less than 6 years of difference to the time period of the data set	Less than 10 years of difference to the time period of the data set	Less than 15 years of difference to the time period of the data set	Age of data unknown or more than 15 years of difference to the time period of the data set
Geographical correlation	Data from area under study	Average data from larger area in which the area under study is included	Data from area with similar production conditions	Data from area with slightly similar production conditions	Data from unknown or distinctly different area (North America instead of Middle East, OECD-Europe instead of Russia)
Further technological correlation	Data from enterprises, processes and materials under study	Data from processes and materials under study (i.e., identical technology) but from different enterprises	Data from processes and materials under study but from different technology	Data on related processes or materials	Data on related processes on laboratory scale or from different technology

[1] Each cell in the matrix indicates a quality characteristic of inventory data. After the analyst has selected, for each item of the inventory, an appropriate cell during the Monte Carlo procedure, the software keeps track of such choice indicating the position (1 to 5) of the selected quality characteristics in each R, C, T, G, F line of the matrix itself.

After a score was assigned to each data indicator for all material and energy inputs included in the inventory, the Ecoinvent data quality system calculates a corresponding numerical value of uncertainty, assigning a specific geometric standard deviation to a log-normal distribution (see the Supplementary Materials for standard deviation values). The propagation of uncertainty throughout the model was then calculated by means of the Monte Carlo analysis (i.e., 10,000 runs), obtaining a final standard deviation on the results in each impact category.

2.2.5. Important Assumptions

Transport of assembled machinery to the geothermal plant site was excluded, because of the limited distance between the plant and the production site (i.e., 150 km). However, the transport of the semi-products and raw material was included using the background processes in the Ecoinvent database.

For small steel parts, an aggregated mass value was provided by EGP. This quantity is supposed to cover all the steel used for general parts in the commissioning phase. Thus, it was equally divided between the six components: AMIS, gas intercooler, gas compressor, condenser, evaporative tower and turbine.

During the maintenance of the power plant, 10% of the steel content of the steam turbine rotor was assumed to be substituted with new steel every four years.

Drilling wastes spent mineral oil and sorbent were considered to be sent to landfill. According to the information supplied, no additional treatment processes were considered.

Data on direct emissions from the power plant's stack was taken from Ferrara et al. [44]. These emissions were modelled as output flows "emission to air, low population density".

2.2.6. Allocation Procedure

The Bagnore 3 and 4 power plant is a multifunctional system, since it produces both electricity and thermal energy. In this study, an exergy-based allocation procedure was chosen to deal with such multifunctionality as a proper allocation method, according to the ILCD Handbook [41]. The exergy allocation method accounts for the quality (i.e., exergy content, ability to do work) of the two energy products (i.e., electricity and heat) generated by the power plant. Thus, 95% of total impacts were allocated to the electricity produced. The complete procedure to calculate allocation coefficients for electricity and heat is reported in detail in the Supplementary Materials, in the sheet "Allocation".

2.2.7. Life Cycle Impact Assessment Method

The ILCD 2011 Midpoint+ method v1.0.9 was adopted for translating the emissions and resources use into environmental impacts, which were quantified during the inventory phase. The impact categories acidification potential (AC); climate change (CC); freshwater ecotoxicity (EC); freshwater eutrophication (FEP); human toxicity, cancer effects (HTc); human toxicity, non-cancer effects (HTnc); ionizing radiation human Health effect (IRHH); land use (LU); marine eutrophication (MEP); mineral, fossil and ren resource depletion (MFRD); ozone depletion (ODP); particulate matter (PM); photochemical ozone formation (POF); terrestrial eutrophication (TE); water resource depletion (WRD) were included in the analysis. The ionizing radiation E (interim) impact category was excluded, due to its incomplete development [45]. The normalization step was performed by applying the reference values of the "EU27 2010, equal weighting" set. According to the latest development in European guidelines of the ILCD method [45], the discussion related to the toxicity categories was excluded from the analysis. All the calculations and modelling were performed using the open-source software OpenLCA version 1.10 and LCIA package v2.0.4 [46].

3. Life Cycle Inventory Analysis

One main objective of this research was to present the most detailed and accurate life cycle inventory based on primary data for a state-of-the-art flash geothermal power plant. The collection of information was performed with the intent of obtaining the highest level of detail in terms of LCA requirements [41]. The inventory of materials and energy input and output flows was collected for each of the separated components, and based only on primary data. To our knowledge, the data inventory built in the present work represents the first of a kind LCI available in the state-of-the-art literature for geothermal power plants based on flash technology. The resulting inventory is presented in its extended version in the Supplementary Materials (sheet "Inventory").

Currently, the most referred LCI available in the literature concerning flash technology is the one published by Karlsdottir et al. [35]. As much as this inventory is quite comprehensive and detailed, and it has often been employed for geothermal system modelling in LCA studies ([24] and references therein), it fails by not providing primary data and accounting for all the life cycle stages of the energy generation system. The present work aims to provide an improved inventory for the flash technology, which could potentially be used in conjunction with the work by Karlsdottir et al. [20,35] for geothermal system modelling in future LCA studies of geothermal power plants.

Table 3 reports the main differences between the present work (right side) and the work by Karlsdottir et al. [35] (left side). Regarding data accuracy, the inventory presented in this work is entirely based on primary data coming from the EGP Company that has executed the activities. Only a few assumptions were made based on expert knowledge, for instance, concerning power building material requirements. In this case, the primary data used for modelling of Bagnore 4 was also used also for Bagnore 3, as suggested by the power plant operator EGP, employing a scaling factor. Even though this can be considered as an estimation, it is still based on primary data and, more importantly, on the expert judgment of the operator. As a result, when considering data quality, a large part of the indicators scored between 1 and 3 (see Table 2).

Table 3. Main differences between the currently available life cycle inventories for flash technology.

Parameter	Karlsdóttir et al. (2015)	This Work
Data accuracy	Not all the data presents the highest level of accuracy. Sometimes secondary data are employed, or data come from extrapolation of secondary data.	Most of the data comes from primary sources, or data are directly extrapolated by the operator Company.
Data coverage	Most of the Life Cycle Stages are analyzed and reported, but • No data coverage for regular maintenance activities. • Direct emissions to air are only partially accounted	All of the Life Cycle Stages are fully analyzed and reported.
System boundaries	End of life (EoL) processes are not included	EoL processes are included; only heating station building, electric supply machinery and distribution infrastructure are not included in the system boundaries

On the other hand, Karlsdottir et al. [35] includes a higher component's specificity, for example, steel grades are provided, as are mass weight for smaller equipment parts. However, these data are mainly based on secondary data and authors' assumptions. The data coverage featured in this paper is higher, compared to the one in Karlsdottir et al. [35]. Specifically, the present work considers all the regular maintenance activities, for example, lubricating oil substitution and regular maintenance operation of machinery, EoL treatments of wastes and wells closure operations, previously never considered.

In this paper, the same approach used in Parisi et al. [37] was adopted. Such an approach relies on a statistical analysis of all the compounds emitted during power generation from geothermal exploitation. The only difference compared to the work of Parisi et al. [37] is represented by the emission values that have been updated with the most recent ones provided by the regional environmental agency [47].

Table 4 reports the main energy and material inputs for the commissioning phase related to the functional unit. Diesel consumption is primarily associated with the wells drilling process with a specific consumption of about 12 GJ/m. Concerning material use, Portland cement and steel represent the most used materials, accounting for about 70% of the total weight of equipment used in this stage. Portland cement is employed in the casing of wells and power station buildings, whereas steel is partitioned among casing, pipelines and machinery. Depending on the application, different steel grades can be used.

The material input for maintenance activities are reported in Table 5. The maintenance stage represents the planned activities required to keep the power plant in operation. Extraordinary maintenance activities are hence omitted. The maintenance activities that result in the highest material consumption are those related to the substitution of the spent Hg absorber (Selenium), the lubricating oil replacement, as well as the steel and polyvinyl chloride (PVC) replacement for power plant machinery. In this case, a substitution of 10% of the total weight of the steam turbine rotor every four years was considered.

Table 4. Main material and energy inputs employed in the commissioning phase. The cut-off is set at 2% of the total mass, to reduce the number of inputs reported. Complete information can be found in the Supplementary Materials.

Energy Input	Amount	Unit
Diesel for drilling	9.4×10^{-3}	MJ/F.U.
Material input	**Amount**	**Unit**
Excavation	6.7×10^{-7}	m^3/F.U.
Portland cement	8.1×10^{-4}	kg/F.U.
Steel	3.3×10^{-4}	kg/F.U.
Gravel	2.5×10^{-4}	kg/F.U.
Bentonite	9.4×10^{-5}	kg/F.U.
Copper	5.0×10^{-5}	kg/F.U.
Sodium hydroxide	3.5×10^{-5}	kg/F.U.
Aluminium	2.7×10^{-5}	kg/F.U.
Material output	**Amount**	**Unit**
Drilling waste to disposal (EoL)	1.0×10^{-3}	kg/F.U.

Table 5. Main material and energy inputs employed in the maintenance phase. The cut-off is set 2% of the total mass, to reduce the number of inputs reported. Complete information can be found in the Supplementary Materials.

Material Input	Amount	Unit
Lubricating oil	5.5×10^{-6}	kg/F.U.
Selenium	4.1×10^{-6}	kg/F.U.
Pentane	2.8×10^{-6}	kg/F.U.
PVC	8.3×10^{-7}	kg/F.U.
Steel	7.7×10^{-7}	kg/F.U.

The operational stage considers the atmospheric emissions, due to geothermal fluid exploitation and the material input needed by the NH_3 abatement system. As shown in Table 6, the emission of CO_2 and methane (CH_4) dominates the environmental emission profile of the Bagnore power plant system. In contrast, the H_2SO_4 is by far the most used material during the operational phase.

Table 6. Main material input and direct atmospheric emissions from the operational phase. The cut-off is set 2% of the total mass to reduce the number of inputs to be reported. Complete information can be found in the Supplementary Materials.

Material Input	Amount	Unit
H_2SO_4	3.7×10^{-3}	kg/F.U.
Atmospheric Emissions	**Amount**	**Unit**
CO_2	4.1×10^{-1}	kg/F.U.
CH_4	1.2×10^{-2}	kg/F.U.

Table 7 provides information on energy and materials inputs for the decommissioning phase. The assumption is that all the drilled wells will undergo a closure process when the plant runs out its lifetime. This approach was adopted more to test the influence of the EoL processes of wells than to represent a real option. The Bagnore power plant system is managed in a sustainable way, ensuring a constant productivity without depletion of the resource. However, since a lifetime must be set in LCA, this work has considered the unlikely option that the wells will be closed after the given lifespan to account for the EoL process.

Table 7. Main material and energy input employed in the well closure phase. The cut-off is set 2% of the total mass, to reduce the number of inputs reported. Complete information can be found in the Supplementary Materials.

Energy Input	Amount	Unit
Diesel	6.4×10^{-4}	MJ/F.U.

Material Input	Amount	Unit
Portland	1.6×10^{-5}	kg/F.U.
Gravel	3.2×10^{-6}	kg/F.U.

4. Results

4.1. Life Cycle Assessment of the Bagnore Power Plant System

Figure 2 reports the percentage of contribution of commissioning, operational, maintenance, decommissioning and EoL phases of the Bagnore power plant system, to the total impacts for all the categories included in the ILCD method. The potential impacts on the 15 categories that were considered are essentially determined by the commissioning and operational phases, which contribute for more than 90% on the total impacts in each category. In more detail, the operational phase contributes for 80%–90% of the overall potential impacts on the AC, CC, HTnc, MEP, PM, POF and TE categories, and about 70% to WRD. These impacts are mainly linked with direct emissions to the air of NH_3, CO_2 and CH_4. The emission of NH_3 determines the impact on AC (i.e., 96% of the total impact), MEP (84% of the total impact), TE (99% of the total impact) and PM (86% of the total impact). In contrast, the impact on the CC category is shared between the CO_2 (i.e., 57% of the total impact) and CH_4 (i.e., 42% of the total impact) emissions. The 75% of the total impact on POF is determined by CH_4.

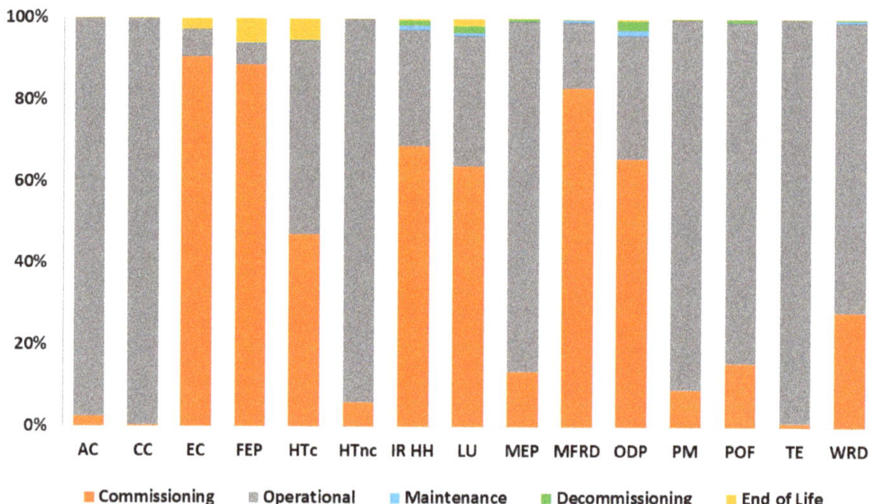

Figure 2. Percentage of contribution of commissioning, operation, maintenance, decommissioning and EoL phases to the total impact in all the assessed impact categories.

The commissioning phase is responsible for more than 80% of the total potential impacts on EC, FEP and MFRD. The copper requirement during the building construction process is the main contributor to such impacts. The commissioning phase contributes about 60% to 70% to the IRHH, LU and ODP categories, for which the deep well construction process shows the highest contribution.

Subsequently, the decommissioning and EoL phases show a negligible contribution to all the considered impact categories.

The characterized results of each impact category were divided by a selected reference value, in order to better understand the magnitude of the results of impact category indicators, and to bring all the results on the same normalized scale (see the Supplementary Materials for normalization values). After the normalization step, the CC, TE and AC categories, in this order, had the highest impacts among all the selected impact categories (Figure 3). The impacts from geothermal electricity production were compared with the impacts from the average Italian energy mix [48], to give a reference system and interpret the magnitude of the geothermal eco-profile. The Ecoinvent version 3.5 employed for the analysis is based on the Italian electric energy mix by 2014. The share consisted of 60% arising from fossils (coal, gas, oil) and import (mostly nuclear). RES represents 40% of the total, with 18% generated by hydro, 7% by photovoltaics, 5% wind, 6% biofuel, 2% waste and 2% geothermal.

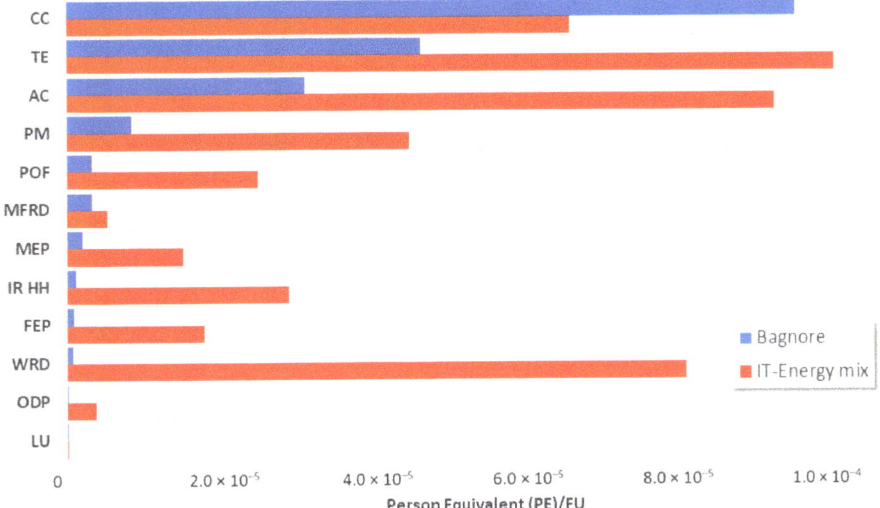

Figure 3. Normalized results for the production of 1 kWh of electric energy from the Bagnore power plant system (blue) and from the average Italian electricity mix (light grey). Climate change (CC), terrestrial eutrophication (TE), acidification potential (AC) and particulate matter (PM) have been identified as the categories with the highest impact.

As shown in Figure 3, all the impacts caused by the average Italian electricity mix are higher than those of geothermal energy production, with the exception of climate change, due to the emissions of CO_2 and CH_4 that are intrinsic to the geothermal resource exploitation activities.

The impacts on the CC, TE and AC categories for geothermal electricity production are determined almost exclusively by emissions to the air during the operational phase (i.e., NH_3, CO_2 and CH_4). As shown in Figure 3 and Table 8, all the impacts caused by the average operational phase are mainly related to the geothermal fluid composition, and can thus be considered site-dependent.

On the contrary, the commissioning phase is common to all flash technologies, and Figure 4 shows the contribution of processes within this phase. The processes considered in the commissioning phase are clustered in drilling, drilling waste (disposal), equipment and pipelines. The drilling process includes, in addition to the drilling activities themselves, the construction of the well pads. In contrast, equipment includes all the materials and energy needed to realize the components present in the power plants and the power plants building itself. The pipelines construction process is separated from the others, because they are structures connecting wells and power plants.

Table 8. Contribution of LC phases to the most impacting categories: AC, CC, PM and terrestrial eutrophication (TE). Impacts are reported as person equivalent (PE) per functional unit (FU). Complete information can be found in the Supplementary Materials.

LC Phase	Unit	AC	CC	PM	TE
Commissioning	PE/FU	8.0×10^{-7}	4.3×10^{-7}	8.0×10^{-7}	5.4×10^{-7}
Operational	PE/FU	3.1×10^{-5}	9.6×10^{-5}	7.8×10^{-6}	4.6×10^{-5}
End of Life	PE/FU	1.1×10^{-9}	2.5×10^{-9}	1.4×10^{-9}	1.3×10^{-9}
Maintenance	PE/FU	7.8×10^{-9}	3.4×10^{-9}	7.0×10^{-9}	2.2×10^{-9}
Decommissioning	PE/FU	1.4×10^{-8}	9.4×10^{-9}	1.1×10^{-8}	2.3×10^{-8}
Total	PE/FU	3.2×10^{-5}	9.6×10^{-5}	8.6×10^{-6}	4.7×10^{-5}

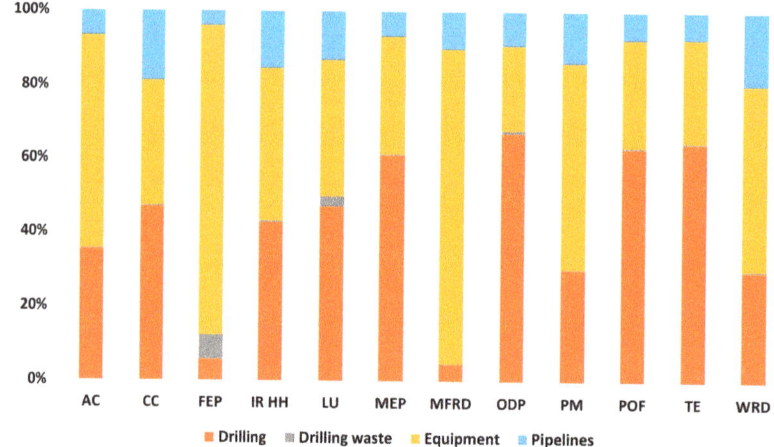

Figure 4. Percentage of contribution from drilling, drilling waste disposal, equipment and pipelines to the commissioning phase.

The hotspot analysis results show that the potential impacts of the commissioning phase are fairly divided among equipment and drilling processes. Building construction and the production of metals (i.e., copper) determine the impact of the equipment. Emissions from the combustion of diesel used to drive the drilling rig are the most responsible for the impact during drilling. Pipelines generally give a contribution of around 10% of the total impacts in all categories, except for CC and WRD. Drilling waste disposal has a negligible impact.

4.2. Uncertainty Analysis of Results

Figure 5 reports the uncertainty values (MIN, MAX and standard deviation) related to the average potential impact for the categories CC, TE, AC and PM, which were previously identified as having the highest impact. The uncertainty associated with the results was calculated following the procedure described in Section 2.2.4.

The calculated uncertainty of results for the identified categories is low and ranging between 2–3%. The impact of these categories is exclusively determined by airborne emissions (primary data) during the operational phase. The good quality of data for airborne emissions (low score in all indicators, see Table 2) results in a low uncertainty of the final LCA results. In those cases where the impact is determined by other stages, with different levels of quality of primary data, the final uncertainty is generally higher and, in some cases, up to a standard deviation around the mean value of 58%.

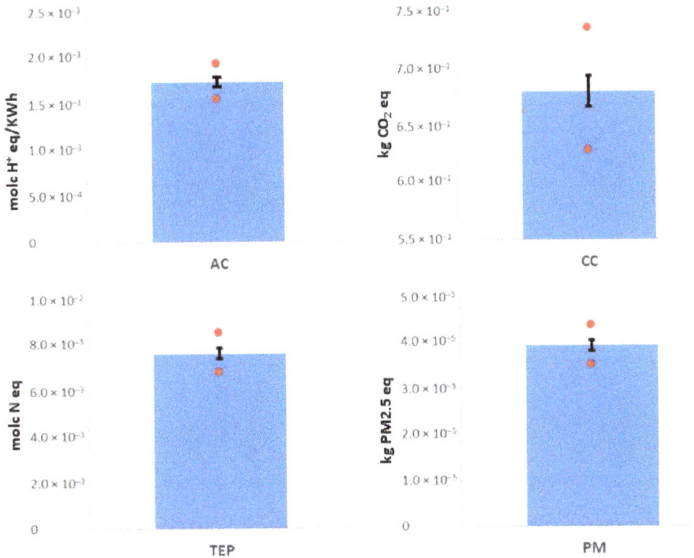

Figure 5. Characterized impact results per kWh of electricity produced for the categories CC, TE, AC and PM. Bars represent the standard deviation around the average impact values, whereas red dots refer to MIN and MAX values.

In Table 9, the uncertainty related to the impacts for all categories is reported together with the overall score for each data quality indicator used to calculate the impacts. Generally, a low overall data quality (high scores in Table 2) corresponds to a relatively high standard deviation (>10%). The uncertainty of results is not exclusively related to the inventory data, but also to the secondary (background) data and their relative uncertainty as specified in the Ecoinvent database.

Table 9. Uncertainty analysis for each impact category results and data quality indicator score. (R) Reliability; (C) Completeness; (T) Temporal correlation; (G) Geographical correlation; (F) Further technological correlation.

Impact Category	Impact Result	STDV (%)	Overall Data Quality Indicator [1]				
			R	C	T	G	F
Acidification (molc H⁺ eq) [2]	1.78×10^{-3}	3	1	1	2	1	1
Climate change (kg CO$_2$ eq)	6.82×10^{-1}	2	1	1	2	1	1
Freshwater eutrophication (kg P eq)	5.19×10^{-6}	58	1	1	4	1	1
Ionizing radiation HH (kBq ^{235}U eq)	2.96×10^{-4}	12	1	1	5	2	1
Land use (kg C deficit)	8.37×10^{-3}	18	3	4	5	3	1
Marine eutrophication (kg N eq)	6.16×10^{-5}	3	1	1	3	1	1
Mineral, fossil & ren resource depletion (kg Sb eq)	6.49×10^{-7}	20	3	2	5	2	2
Ozone depletion (kg CFC-11 eq)	4.39×10^{-10}	47	2	3	4	5	3
Particulate matter (kg PM2.5 eq)	4.35×10^{-5}	3	1	1	2	1	1
Photochemical ozone formation (kg NMVOC eq)	1.52×10^{-4}	3	1	1	3	1	1
Terrestrial eutrophication (molc N eq) [2]	7.69×10^{-3}	3	1	1	2	1	1
Water resource depletion (m^3 water eq)	5.69×10^{-5}	11	2	3	5	3	2

[1] numbers in columns R, C, T, G, F refer to specific scores within the EcoInvent uncertainty matrix (Table 2). [2] molc unit indicates a mole of charge (molc) per unit of mass emitted.

5. Discussion

The LCA results show that direct emissions to the atmosphere released during the commissioning and operational phases are the dominant impact for the Bagnore system. For the commissioning phase, as the emissions of CO$_2$ associated with the combustion of diesel used to drive the drilling rig are the principal factors responsible for the environmental impact, the eco-profile would certainly improve in the future by changing the drilling technology. Unfortunately, so far, the initiatives promoted by operators to employ an electric rig, directly powered by the medium-voltage network, aiming for a simplification of the process and a reduction of impact and costs, have been unsuccessful. The main difficulties arose with the medium-voltage network connections, and for authorization procedures, which look quite complex due to safety requirements.

However, the applicability of such a system looks only suitable for the consolidated stations with several wells.

The potential environmental impacts generated during the operational phase are mainly linked with airborne emissions. The comparison with the Italian energy mix makes it possible to highlight the differences in the environmental performances, which are in favor of geothermal energy exploitation for all the environmental impact categories, with the exception of climate change. This outcome is due to the significant contribution given to the average Italian electricity mix from RES like hydro, photovoltaics and wind energy, whose CO_2 emission contributions in the atmosphere during the operational phase are negligible. This confirms previous evidence that geothermal energy, although renewable, is not the cleanest one, even if it performs better than any other fossil source. This finding gives a benchmark to interpret the magnitude of the power plant eco-profile. As the emissions of NH_3, CH_4 and CO_2 during the operational phase are mainly related to the geothermal fluid composition, they can be considered site-dependent, therefore particular care should be exercised in deciding the localization of plants in the project phase. In this context, it should be mentioned that, to date, in the analysis of greenhouse gases emissions, the Intergovernmental Panel on Climate Change [49] considers the release of greenhouse gases of geothermal origin quantitatively negligible, despite the fact that this has been demonstrated not always to be true [7]. Notwithstanding the evidence that flash geothermal electricity production is contributing to CO_2 emissions more than the Italian electricity mix, some intrinsic benefits connected with geothermal development should be considered. This is particularly important in the frame of a policy sensitive to environmental and social issues: (i) geothermal energy is a renewable local based energy source and not imported; (ii) a secondary, but not negligible advantage can be found in the use of thermal fluids for civil or light industry purposes in the neighboring area; (iii) electricity generated by geothermal contributes to the basic load and it is independent on atmospheric conditions. However, regarding this latter issue, we should be aware that in the future, due to discontinuity of solar and wind electricity supply, flexible power systems will be even more valuable.

The main achievement of the assessment method implemented in this work relies mainly on two aspects. Firstly, the investigated system has been selected from the latest in technological excellence in the field of flash geothermal generation in Italy. Secondly, the EGP operator granted the availability of primary data to build the LCI, as reported in the Supplementary Materials. This is noteworthy compared to the state-of-the-art LCA literature on geothermal systems, which very often uses secondary or tertiary data.

The representativeness and quality of the inventory data, presented in Section 3, should always be assessed to ensure robustness of LCA results. Significant elements of improvement in this work are represented by the level of detail for machinery and components, data quality and coverage, as well as the inclusion of the EoL as shown in Table 3.

The exergy-based allocation method chosen to address the multifunctionality represents another feature of this work: although not fully new, most LCA studies allocate according to mass, energy content or monetary value. Exergy reflects the difference in terms of energy quality among energy outputs and represents the most suitable method, from a thermodynamic point of view, for discerning the benefits of combined heat and power systems.

The uncertainty evaluation on LCA results performed with the Monte Carlo analysis shows a non-negligible dependence on the background Ecoinvent database and the LCIA method assumptions, not on foreground data. This confirms the reliability of the LCA system modelling adopted in this work. The scientific approach employed offers a detailed insight of the research findings in agreement with the ISO 14040 and ILCD requirements. From a policy point of view, the transparency of the assessment method could support effective decision-making.

As mentioned in the Introduction, the EU has committed itself to a clean energy transition, which will contribute to fulfilling the goals of the Paris Agreement on climate change [1,5]. According to the Italian NECP targets [6], the electric generation power will be affected by an important transformation, due to the goal of the phasing out of coal generation plants by 2025, and the necessary promotion of a large contribution from RES to replace them. The maximum contribution to the growth of RES will arise particularly from the electricity sector, which, by 2030, will reach 187 TWh of generation from RES, equal to 16 Mtep. The strong penetration of technologies for renewable electrical energy production, mainly photovoltaics and wind, will make it possible to cover 55.0% of the final gross electric consumptions with RES, compared to the 40% contributed in 2014. The photovoltaics and wind capacity should triple and double, respectively, by 2030. With regards to other RES in the NECP, a limited growth of additional geothermal power from 813 to 950 MWe is foreseen, which would represent the only maintenance of the actual 2% of the Italian electric mix. This target is considered for conventional geothermal technology, with reduced direct emission limits. It arises from the awareness that, even if geothermal energy is quite suitable for replacing fossils in electricity production, the limits due to environmental impacts still hold. The possibility of providing incentives for other technologies like that with zero emissions in plants, with a total reinjection of fluids, is under evaluation. At the moment, these technologies, like geothermal at reduced environmental impact, are considered to be as innovative in the national context as offshore wind, concentrated solar power and ocean energy.

6. Conclusions

In this paper, a cradle to grave LCA of the Italian flash technology Bagnore power plants system has been performed, based on a comprehensive and accurate life cycle inventory of primary data supplied by the plant manufacturer and operator EGP. From the LCA results, it can be inferred that the potential environmental impacts are determined for more than 95% by the operational (direct emissions to air of NH_3, CH_4, CO_2) and commissioning (CO_2 emissions due to diesel combustion during drilling) phases. Maintenance, decommissioning and EoL phases show a negligible contribution to all the considered impact categories. Globally, out of the sixteen impact categories selected, climate change, ccidification, terrestrial eutrophication and particulate matter were the most affected. These outcomes imply that LCA results of electricity generation from flash technology employing a mid to high dissolved gas content fluid are primarily determined by emissions to air. Direct emissions into the atmosphere are the responsible for most of the environmental impact in the operational phase (84%). The comparison made with the life-cycle environmental impacts caused by the production process of the average Italian electricity mix showed that the balance is almost always in favor of the geothermal energy production, with the only exception being the climate change category. A further finding of this work is that, in the commissioning phase, the impact is equally divided between well drilling and equipment. It is notable that the copper requirement during the building construction process is the main contributor to impact in the commissioning phase. Accordingly, future research might explore the possibility of replacing metals and particularly copper in building the plant.

It should be noticed that the data referring to the commissioning, maintenance and EoL stages presented in this study might be used by the scientific community to evaluate the potential environmental impact of geothermal systems. On the other hand, site-specific information, such as direct environmental emissions measured during the operational phase, is exclusively valid in this specific geothermal field.

This work offers the most complete life cycle inventory for a state-of-the-art flash system conversion technology accounting for the whole life cycle of the geothermal power plant. The robustness of the results obtained here, as demonstrated by the uncertainty analysis, emphasises the need for high quality primary data for performing reliable and consistent LCA studies. This is particularly true in the geothermal sector, where the lack of primary data and precise information about the conversion technology and the geo-specificity of the reservoir for long periods prevented the possibility to get reliable results, affecting the quality of the LCA studies. In this context, the availability of primary data and open access to technical repositories become essential to reach high standards in the LCA literature concerning geothermal systems. We believe that the accurate approach presented in this paper will aid promoting the implementation of environmental assessment studies, which are essential to undertake impact minimization actions on currently operating power plants, and to improve the eco-design perspective of future installations.

Supplementary Materials: The following are available online at http://www.mdpi.com/1996-1073/13/11/2839/s1, Excel file S1: energies-819511-supplementary.xls.

Author Contributions: Conceptualization, R.B. and M.L.P.; methodology, R.B. and M.L.P.; validation, R.B. and M.L.P.; formal analysis, L.T., N.F. and M.L.P.; investigation, L.T., N.F. and M.L.P.; data curation, L.T., N.F. and M.L.P.; writing-original draft preparation, L.T., N.F., R.B. and M.L.P.; writing-review and editing, L.T., N.F., R.B. and M.L.P.; visualization, L.T. and N.F.; supervision, R.B. and M.L.P.; project administration, R.B. and M.L.P.; funding acquisition, M.L.P. All authors have read and agreed to the published version of the manuscript.

Funding: This research received no external funding.

Acknowledgments: MIUR Grant—Department of Excellence 2018–2022. L.T. and M.L.P. acknowledge the European Union's Horizon 2020 Framework Program for funding Research and Innovation under Grant agreement no. 818242 (GEOENVI) for funding. Authors would like to thank Romina Taccone, Marco Paci and Massimo Luchini (EGP) for primary data supplying. EGP and Roberto Bonciani are gratefully acknowledged for fruitful discussion. NF thanks COSVIG (Regione Toscana) for contributing to his grant. Careful reading and revising of the manuscript by Emeritus Michael Rodgers, Bowling Green State University, is gratefully acknowledged.

Conflicts of Interest: The authors declare no conflict of interest.

Abbreviations

AMIS	Atmospheric emissions abatement system
AC	Acidification potential
CC	Climate change
EC	Freshwater ecotoxicity
EGP	Enel Green Power
EGS	Enhanced Geothermal System
EoL	End of Life
FEP	Freshwater eutrophication
GHG	Greenhouse gas
GWh/y	Giga Watt hour per year
HTc	Human toxicity, cancer effects
HTnc	Human toxicity, non-cancer effects
IRE	Ionizing radiation Environment effect (interim)
IRHH	Ionizing radiation Human Health effect
LCA	Life cycle assessment
LCI	Life cycle inventory
LCIA	Life cycle impact assessment
LU	Land use
MEP	Marine eutrophication
MFRD	Mineral, fossil & ren resource depletion
MWe	Mega Watt electric
NCGs	Non-condensable gasses
ODP	Ozone depletion
ORC	Organic Rankine cycle
PM	Particulate matter
POF	Photochemical ozone formation
RES	Renewable energy sources
TE	Terrestrial eutrophication
WRD	Water resource depletion

References

1. European Commission. *Clean Energy for all Europeans*; European Commission: Brussels, Belgium, 2019.
2. EU. Directive (EU) 2018/2001 of the European Parliament and of the Council on the promotion of the use of energy from renewable sources. *Off. J. Eur. Union* **2018**, *2018*, 82–209.
3. European Commission. *EUR-Lex—52014DC0015—EN 2014*; European Commission: Brussels, Belgium, 2014.
4. European Commission. *A Clean Planet for All A European Strategic Long-Term Vision for a Prosperous, Modern, Competitive and Climate Neutral Economy*; European Commission: Brussels, Belgium, 2018; pp. 1–25.
5. European Commission. *EUR-Lex—32018R0842—EN 2018*; European Commission: Brussels, Belgium, 2018.
6. European Commission. *EUR-Lex—52018DC0716—EN 2018*; European Commission: Brussels, Belgium, 2018.
7. Bravi, M.; Basosi, R. Environmental impact of electricity from selected geothermal power plants in Italy. *J. Clean. Prod.* **2014**, *66*, 301–308. [CrossRef]
8. Dumas, P.; Garabetian, T.; Serrani, C.; Pinzuti, V. *EGEC Geothermal Market Report 2018*; EGEC: Brussels, Belgium, 2019.
9. Fridleifsson, I.B.; Bertani, R.; Huenges, E. The possible role and contribution of geothermal energy to the mitigation of climate change. In *Proceedings of the IPCC Scoping Meeting on Renewable Energy Sources*; Hohmeyer, O., Trittin, T., Eds.; The Intergovernmental Panel on Climate Change: Luebeck, Germany, 2008; pp. 59–80.
10. International Energy Agency. *Geothermal—Tracking Power*; International Energy Agency: Paris, France, 2019.
11. International Energy Agency. *World Energy Outlook*; International Energy Agency: Paris, France, 2019.
12. Bertani, R. Geothermal power generation in the world 2010–2014 update report. *Geothermics* **2016**, *60*, 31–43. [CrossRef]
13. Sanner, B. Summary of EGC 2019 Country Update Reports on Geothermal Energy in Europe. In Proceedings of the European Geothermal Congress 2019, The Hague, The Netherlands, 11–14 June 2019; p. 18.
14. Turconi, R.; Boldrin, A.; Astrup, T. Life cycle assessment (LCA) of electricity generation technologies: Overview, comparability and limitations. *Renew. Sustain. Energy Rev.* **2013**, *28*, 555–565. [CrossRef]
15. Parisi, M.L.; Maranghi, S.; Vesce, L.; Sinicropi, A.; Di Carlo, A.; Basosi, R. Prospective life cycle assessment of third-generation photovoltaics at the pre-industrial scale: A long-term scenario approach. *Renew. Sustain. Energy Rev.* **2020**, *121*, 109703. [CrossRef]
16. Rossi, F.; Parisi, M.L.; Maranghi, S.; Basosi, R.; Sinicropi, A. Environmental analysis of a Nano-Grid: A Life Cycle Assessment. *Sci. Total Environ.* **2019**, *700*, 134814. [CrossRef]
17. Basosi, R.; Bonciani, R.; Frosali, D.; Manfrida, G.; Parisi, M.L.; Sansone, F. Life cycle analysis of a geothermal power plant: Comparison of the environmental performance with other renewable energy systems. *Sustainability* **2020**, *12*, 2786. [CrossRef]
18. Buonocore, E.; Vanoli, L.; Carotenuto, A.; Ulgiati, S. Integrating life cycle assessment and emergy synthesis for the evaluation of a dry steam geothermal power plant in Italy. *Energy* **2015**, *86*, 476–487. [CrossRef]
19. Parisi, M.L.; Basosi, R. Geothermal energy production in Italy: An LCA approach for environmental performance optimization. In *Life Cycle Assessment of Energy Systems and Sustainable Energy Technologies—The Italian Experience*; Basosi, R., Cellura, M., Longo, S., Parisi, M.L., Eds.; Springer: Berlin/Heidelberg, Germany, 2019; pp. 31–43. ISBN 18653529. [CrossRef]
20. Karlsdottir, M.R.; Heinonen, J.; Palsson, H.; Palsson, O.P. Life cycle assessment of a geothermal combined heat and power plant based on high temperature utilization. *Geothermics* **2020**, *84*, 101727. [CrossRef]
21. Paulillo, A.; Striolo, A.; Lettieri, P. The environmental impacts and the carbon intensity of geothermal energy: A case study on the Hellisheiði plant. *Environ. Int.* **2019**, *133*, 105226. [CrossRef]
22. Bayer, P.; Rybach, L.; Blum, P.; Brauchler, R. Review on life cycle environmental effects of geothermal power generation. *Renew. Sustain. Energy Rev.* **2013**, *26*, 446–463. [CrossRef]
23. Menberg, K.; Pfister, S.; Blum, P.; Bayer, P. A matter of meters: State of the art in the life cycle assessment of enhanced geothermal systems. *Energy Environ. Sci.* **2016**, *9*, 2720–2743. [CrossRef]
24. Tomasini-Montenegro, C.; Santoyo-Castelazo, E.; Gujba, H.; Romero, R.J.; Santoyo, E. Life cycle assessment of geothermal power generation technologies: An updated review. *Appl. Therm. Eng.* **2017**, *114*, 1119–1136. [CrossRef]

25. Asdrubali, F.; Baldinelli, G.; D'Alessandro, F.; Scrucca, F. Life cycle assessment of electricity production from renewable energies: Review and results harmonization. *Renew. Sustain. Energy Rev.* **2015**, *42*, 1113–1122. [CrossRef]

26. Sullivan, J.L.; Clark, C.E.; Han, J.; Wang, M. Life cycle analysis of geothermal systems in comparison to other power systems. *Trans. Geotherm. Resour. Counc.* **2010**, *34*, 128–132.

27. Sullivan, J.L.; Clark, C.E.; Yuan, L.; Han, J.; Wang, M. *Life-Cycle Analysis Results for Geothermal Systems in Comparison to Other Power Systems: Part. II*; Argonne: Lemont, IL, USA, 2012.

28. Lacirignola, M.; Meany, B.H.; Padey, P.; Blanc, I. A simplified model for the estimation of life-cycle greenhouse gas emissions of enhanced geothermal systems. *Geotherm. Energy* **2014**, *2*, 1–19. [CrossRef]

29. Lacirignola, M.; Blanc, P.; Girard, R.; Pérez-López, P.; Blanc, I. LCA of emerging technologies: Addressing high uncertainty on inputs' variability when performing global sensitivity analysis. *Sci. Total Environ.* **2017**, *578*, 268–280. [CrossRef] [PubMed]

30. Marchand, M.; Blanc, I.; Marquand, A.; Beylot, A.; Bezelgues-Courtade, S.; Traineau, H. Life Cycle Assessment of High Temperature Geothermal Energy Systems. In Proceedings of the World Geothermal Congress 2015, Melbourne, Australia, 19–24 April 2015; pp. 19–25.

31. Martínez-Corona, J.I.; Gibon, T.; Hertwich, E.G.; Parra-Saldívar, R. Hybrid life cycle assessment of a geothermal plant: From physical to monetary inventory accounting. *J. Clean. Prod.* **2017**, *142*, 2509–2523. [CrossRef]

32. Frick, S.; Kaltschmitt, M.; Schröder, G. Life cycle assessment of geothermal binary power plants using enhanced low-temperature reservoirs. *Energy* **2010**, *35*, 2281–2294. [CrossRef]

33. Pratiwi, A.; Ravier, G.; Genter, A. Geothermics Life-cycle climate-change impact assessment of enhanced geothermal system plants in the Upper Rhine Valley. *Geothermics* **2018**, *75*, 26–39. [CrossRef]

34. European Commission. *SET Plan Integrated SET—Plan Action 7 "Become Competitive in the Global Battery Sector to Drive E-Mobility and Stationary Storage Forward"*; European Commission: Brussels, Belgium, 2018; pp. 1–70.

35. Karlsdóttir, M.R.; Pálsson, Ó.P.; Pálsson, H.; Maya-Drysdale, L. Life cycle inventory of a flash geothermal combined heat and power plant located in Iceland. *Int. J. Life Cycle Assess.* **2015**, *20*, 503–519. [CrossRef]

36. Fedeli, M.; Mannari, M.; Sansone, F. BAGNORE 4: A benchmark for geothermal power plant environmental compliance. In Proceedings of the European Geothermal Congress, Strasbourg, France, 19–23 September 2016.

37. Parisi, M.L.; Ferrara, N.; Torsello, L.; Basosi, R. Life cycle assessment of atmospheric emission profiles of the Italian geothermal power plants. *J. Clean. Prod.* **2019**, *234*, 881–894. [CrossRef]

38. Bonciani, R.; Lenzi, A.; Luperini, F.; Sabatelli, F. Geothermal power plants in Italy: Increasing the environmental compliance. In Proceedings of the Conference European Geothermal Congress 2013, Pisa, Italy, 3–7 June 2013.

39. International Organization for Standardization. *ISO 14040:2006—Environmental Management—Life Cycle Assessment—Principles and Framework 2006*; International Organization for Standardization: Geneva, Switzerland, 2006.

40. International Organization for Standardization. *ISO 14044:2006 Environmental management—Life cycle assessment—Requirements and guidelines 2006*; International Organization for Standardization: Geneva, Switzerland, 2006.

41. European Commission. *International Reference Life Cycle Data System (ILCD) handbook: Framework and Requirements for Life Cycle Impact Assessment Models and Indicators*; European Commission: Brussels, Belgium, 2010; ISBN 9789279175398.

42. Wernet, G.; Bauer, C.; Steubing, B.; Reinhard, J.; Moreno-Ruiz, E.; Weidema, B. The ecoinvent database version 3 (part I): Overview and methodology. *Int. J. Life Cycle Assess.* **2016**, *21*, 1218–1230. [CrossRef]

43. Weidema, B.P.; Bauer, C.; Hischier, R.; Mutel, C.; Nemecek, T.; Reinhard, J.; Vadenbo, C.O.; Wenet, G. *Overview and Methodology. Data Quality Guideline for the Ecoinvent Database Version 3. Ecoinvent Report 1 (v3)*; The ecoinvent Centre: St. Gallen, Switzerland, 2013.

44. Ferrara, N.; Basosi, R.; Parisi, M.L. Data analysis of atmospheric emission from geothermal power plants in Italy. *Data Br.* **2019**, *25*, 104339. [CrossRef] [PubMed]

45. Saouter, E.; Biganzoli, F.; Ceriani, L.; Versteeg, D.; Crenna, E.; Zampori, L.; Sala, S.; Pant, R. *Environmental Footprint: Update of Life Cycle Impact Assessment Methods: Ecotoxicity Freshwater, Human Toxicity Cancer, and Non-Cancer*; Publications Office of the European Union: Luxembourg, 2018.

46. Greendelta OpenLCA V 1.8 2018. Available online: http://www.openlca.org/greendelta/ (accessed on 11 January 2020).
47. ARPAT. *Geothermal Emissions Monitoring*; Tuscany Regional Agency for Environmental Protection: Pisa, Italy, 2018. (In Italian)
48. Itten, R.; Frischknecht, R.; Stucki, M.; Scherrer, P.; Psi, I. *Life Cycle Inventories of Electricity Mixes and Grid*; Paul Scherrer Institute: Villigen, Switzerland, 2014; pp. 1–229.
49. Stocker, T.F.; Qin, D.; Plattner, G.-K.; Tignor, M.; Allen, S.K.; Boschung, J.; Nauels, A.; Xia, Y. *Climate Change 2013: The Physical Science Basis. Contribution of Working Group I to the Fifth Assessment Report of the Intergovernmental Panel on Climate Change*; IPCC: Geneva, Switzerland, 2013.

Article

Environmental and Comparative Assessment of Integrated Gasification Gas Cycle with CaO Looping and CO$_2$ Adsorption by Activated Carbon: A Case Study of the Czech Republic

Kristína Zakuciová [1,2,*], Ana Carvalho [3], Jiří Štefanica [1], Monika Vitvarová [4], Lukáš Pilař [4] and Vladimír Kočí [2,*]

[1] ÚJV Řež, a. s., Hlavní 130, Řež, 250 68 Husinec, Czech Republic; jiri.stefanica@ujv.cz

[2] Department of Environmental Chemistry, Faculty of Environmental Technology, University of Chemistry and Technology, Prague, Technická 5, 166 28 Praha 6, Czech Republic

[3] CEG-IST, Instituto Superior Técnico, Universidade de Lisboa, 1049-001 Lisbon, Portugal; anacarvalho@tecnico.ulisboa.pt

[4] Department of Energy Engineering, Faculty of Mechanical Engineering, Czech Technical University in Prague, Technická 4, 166 28 Praha 6, Czech Republic; Monika.Vitvarova@fs.cvut.cz (M.V.); Lukas.Pilar@fs.cvut.cz (L.P.)

* Correspondence: kristina.zakuciova@ujv.cz (K.Z.); vlad.koci@vscht.cz (V.K.); Tel.: +420-775-364-032 (K.Z.)

Received: 30 June 2020; Accepted: 12 August 2020; Published: 13 August 2020

Abstract: The Czech Republic is gradually shifting toward a low-carbon economy. The transition process requires measures that will help to contain energy production and help to reduce emissions from the coal industry. Viable measures are seen in carbon capture technologies (CCTs). The main focus is on the environmental and economic comparison of two innovative CCTs that are integrated in the operational Czech energy units. The assessed scenarios are (1) the scenario of pre-combustion CO$_2$ capture integrated into the gasification combined cycle (IGCC-CaL) and (2) the scenario of post-combustion capture by adsorption of CO$_2$ by activated carbon (PCC-A). An environmental assessment is performed through a life-cycle assessment method and compares the systems in the phase of characterization, normalization, and relative contribution of the processes to the environmental categories. Economic assessment compares CCT via capture and avoided costs of CO$_2$ and their correlation with CO$_2$ allowance market trend. The paper concludes with the selection of the most suitable CCT in the conditions of the Czech Republic by combining the scores of environmental and economic parameters. While the specific case of IGCC-CaL shows improvement in the environmental assessment, the economic analysis resulted in favor of PCC-A. The lower environmental–economic combination score results in the selection of IGCC-CaL as the more viable option in comparison with PCC-A in the current Czech energy and economic conditions.

Keywords: carbon dioxide capture; activated carbon; environmental impacts; IGCC; carbon capture economy

---

1. Introduction

Energy self-sufficiency and low-carbon-economy transition are the concepts currently forcing the coal-based energy industry to significantly decrease its produced emissions. The annual consumption of coal (in coal-based industries) in the Czech Republic is over 60 million tons per year, ranking Czech Republic at 17th in worldwide consumption [1]. The Czech energy industry is also dependent on imports of 97% of oil and gas [1]. Several measures were adopted to reduce energy import dependence, including implementation of a higher share of renewable energy sources and more efficient use of fossil fuels such as brown coal [1]. However, the transition process cannot be sudden and must be gradually implemented by viable investments. One such investment, which seems to be feasible for the current Czech energy industry but also usable for steel industries, is carbon capture technologies (CCTs). These technologies must be carefully assessed and planned for the specific conditions of a given country. There are three parameters that must be considered for the feasibility assessment of CCT—(i) technological feasibility, (ii) economic performance, and (iii) environmental performance. This paper considers each of these three parameters in a new combined analysis.

In the conditions of the Czech operational power units, several CCT options were considered, such as post-combustion technologies of ammonia scrubbing, activated carbon adsorption (PCC-A), and pre-combustion integrated gasification gas cycle with integrated carbonate loop (IGCC-CaL). These technologies are the subject of intensive research and optimization to achieve their implementation into the operational power units. The decision-making process for the choice of suitable technologies may be significantly influenced by environmental performance consideration via comprehensive methodology. Life cycle assessment (LCA) is one of the best certified methods to create environmental models of the considered systems. LCA allows the comparison of the assessed systems among each other [2].

The ammonia scrubbing process of LCA was already performed [3]. The ammonia scrubbing process increased the impact of fossils depletion and mineral resource depletion in comparison with the power unit (250 MWe) without CCT. That is caused by a large amount of additional energy consumption for ammonia solvent preparation. Moreover, energy efficiency of the power unit decreased from 38% to 27%. The environmental problem occurs with the treatment of ammonia salts, currently considered as non-utilized waste. On the other hand, CO_2 was captured in a ratio of 90%.

An LCA for the PCC-A system for Czech conditions was recently published by the authors of this paper [4]. The LCA model in the study considers a functional unit nominal power output of 250 MWe. The paper concludes that adding such technology would increase the energy demand (an additional 1133 MJ for hard coal activation) and fossils depletion. The reason for this is the resource consumption of hard coal (additional fresh carbon 23 kg/h) in the production chain of the activated carbon.

IGCC-CaL was not previously assessed for the Czech conditions from the environmental point of view. However, several studies were made for the IGCC systems integration and its environmental assessment. A summary of the following studies can be found in Table 1.

The extensive study by Singh et al. [5] compares the environmental results for 400 MWe power plants with post, pre, and oxy-fuel combustion capture systems. For pre-combustion systems with IGCC based on Selexol absorption, the CO_2 capture ratio is 90%, with an energy efficiency of 37.6%. Comparative LCA was made by hybrid LCA approach, using input–output analysis together with the ReCiPe 2008 version 1.02 method. Environmental results of a pre-combustion system show a reduction of 78% in the category of global warming potential (GWP), the highest reduction in comparison with the aforementioned systems. On the other hand, IGCC system contributes to increase of 120% in category of freshwater eutrophicaton (FE), influenced mainly by infrastructure development.

Cormos C. [6] evaluated the techno-economic and environmental performance of IGCC system for power plant concepts of a net power output of about 400–500 MWe. The study states that the introduction of the CCT system decreased net plan energy efficiency by 7.1–9.5%. The environmental part of the study compared the IGCC systems based on the physical solvent Selexol. Environmental impacts refer to the production of 1 kWh of electricity. However, the impacts were categorized in normalized mass and energy flows where the integration of the IGCC systems caused an increase of coal (25%), oxygen (24%), and cooling water (22%) consumption, and the ratio of captured CO_2 was modelled at 90%.

Table 1. Review of references focused on environmental assessment of pre-combustion integrated gasification gas cycle technology with CO_2 capture (IGCC-CCT), global warming potential (GWP), acidification potential (AP), fossils depletion (FD), eutrophication potential (EP), terrestrial acidification (TA), and freshwater eutrophication (FE).

Studies	Type/Size/Net Energy Efficiency of Reference Power Plant	Type/Size of IGCC	Type/Size of IGCC-CCT	IGCC Efficiency (without CCT)	IGCC Efficiency with CCT	CO_2 Capture Rate	CO_2 Specific Emissions for CCT	Environmental Assessment Methodology	Main Environmental Results of IGCC + CCT	CO_2 Product Specification
Falcke et al. (2011)	Supercritical boiler with limestone desulphurization—455 Mwe/37%	Oxygen blown IGCC/394 MWe	Entrained-flow gasification, sour water gas shift reactors, use of Selexol—321 MWe	32.1%	26.1%	81%	N/A	Mass-energy flows	Environmental results expressed graphically	Co-sequestration CO_2 with H_2S removal; CO_2 compression to 100 bar, CO_2 dried by triethylen glycol
Singh et al. (2011)	Supercritical coal power plant /400 MWe/43.4%	IGCC power unit + gasification, gas cleaning unit + gas-fired combined cycle/400 MWe	Pre-combustion capture use of Selexol/400 MWe	44.1%	37.60%	90%	85.7 g/kWh	Hybrid LCA input-output model and ReCiPe 2008 v1.02	GWP (1.8×10^{-1} kg CO_2eqv/kWh); TA (1.1×10^{-3} kg SO_2 eqv./kWh) FE (2.3×10^{-6}/kWh)	CO_2 compressed to 110 bar and transported over 500 km
Cormos (2012)	Coal based IGCC	IGCC based on Shell gasifier/ 485 MWe	Pre-combustion capture based on Selexol/433.18 MWe	46.61%	37.11%	90%	86.92 kg/MWh	Normalized mass energy balance method	Environmental results expressed as mass or energy/MWh; Introduction of CCT increase all normalized flows	CO_2 drying by triethylen glycol; CO_2 compression to 110 bar
		IGCC based on Siemens gasifier/ 448.97 MWe	Pre-combustion capture based on Selexol/420.41 MWe	43.13%	36.02%	90%	76.12 kg/MWh			
Petrescu et al. (2017)	IGCC power plant/493.13 MWe	IGCC power plant with acid gas removal based on Selexol	Pre-combustion IGCC-CaO looping/607.82 MWe	45.09%	36.44%	91.56%	58.87 kg/MWh	LCA method based on CML 2001, GaBi software v. 6	GWP (917.25 kg CO_2 eqv/MWh), AP (1.47 kg SO_2 eqv/MWh), EP (1461.97 kg PO_4^{3-} eqv/MWh)	Co-sequestration with hydrogen-rich gas; CO_2 dried and compressed up to 120 bar, transported 800 km
			Pre-combustion IGCC-FeL based oxygen carriers chemical looping/443.07 MWe		38.76%	99.45%	3.01 kg/MWh		GWP (338.73 kg CO_2 eqv/MWh), AP (1.75 kg SO_2 eqv/MWh), EP (1949.12 kg PO_4^{3-} kg eqv/MWh)	

In another study [7], three following three systems were compared: (i) a conventional supercritical coal power system, (ii) an IGCC-CCS system based on Selexol solvent system, (iii) and an IGCC without CCS. The systems were chosen according to equal coal consumption rather than equal electricity production. The environmental results for IGCC systems per 1 MWh are based on mass–energy balances and show higher water consumption due to the gasification process and the shift reaction in comparison to the power unit without CCT. The IGCC system with CO_2 capture was designed for 81% CO_2 capture ratio. The IGCC-CCS system reduced net energy efficiency from 32.1% (IGCC without CCS) to 26.1%.

The work of Petrescu et al. [8] compares IGCC power plant with gross electric output of 570.61 MWe and 2 IGCC-CCS systems. Two compared IGCC-CCS systems are based on Ca-based (IGCC-CaL) sorbents and iron-based oxygen carriers (IGCC-FeL). The used method for LCA analysis was CML 2001 using GaBi software. In both scenarios, the highest values refer to GWP, where the majority (85%) comes from coal mining and extraction. The results show that the highest carbon capture rate happens with IGCC-FeL (99.45%), with a net electrical efficiency drop from 45.09% to 38.76%. Energy efficiency dropped for IGCC-CaL from 45.09% to 36.44%, and the capture rate was 91.56%.

Regarding CaO looping, some studies were done for post-combustion CO_2 capture. One study [9] considers 600 MWe supercritical pulverized coal power plant as a basis for CCT. According to the study [9], net energy efficiency drops from 39% to 32% due to CaO looping. LCA analysis was done at the endpoint level via SimaPro v8.3 software. The results indicate an increase in resources depletion, ozone depletion, and toxicities. The climate change impact was reduced by 72%.

Clarens et al. [10] compared a sub-critical coal power plant without CCT (500 MWe; net plant efficiency 36.9%), post-combustion capture technologies based on amine absorption (Econamine and Econamine FG+), and CaO looping without capture. This study used the LCA method of ReCiPe v1.04 and Simapro software. The CaO looping results in the best environmental performance among all systems in the categories of ozone depletion, photochemical ozone formation, particulate matter, and water depletion impacts. The CO_2 capture ratio was 90% in each CCT systems. However, the net plant efficiency for the Econamine case dropped from 36.9% to 22.3%, for Econamine FG+ to 25.9%, and for CaO loop to 29.6%.

The literature survey clearly shows that very few environmental assessments were done on the subject of pre-combustion IGCC-CaL and none for comparison of IGCC-CaL and PCC-A. Moreover, the data in the studies is, in the majority, based on literature sources and heat-mass models rather than real case studies. Also, the selected environmental assessment methods were based on mass-energy flows analysis or methods such as hybrid LCA or CML.

The first part of the paper is focused on the environmental study that compares both CCT systems integrated into their reference power plants. The environmental study does not evaluate the reference power plants without CCT due to lack of data for a single IGCC system. Moreover, IGCC-CaL was designed as the one whole technology with an already integrated CCT system. Yet, in the case of PCC-A, a recent study [4] published by the same authors compares the reference 250 MWe power unit and the same reference power unit with PCC-A.

The second part of the paper is the economic study of both systems. The economic part compares the investments of both systems (IGCC-CaL, PCC-A) with the case of the energy system without CCT based on the cost and market trend of CO_2 allowances.

The third part of the paper combines environmental and economic results to determinate the specifics that can influence the decision-making process for the final technology selection.

The paper has several contributions:

- Environmental performance of two innovative systems in the conditions of the Czech energy mix;
- Comprehensive LCA model including decision-making processes of characterization, normalization, pareto analysis;
- Identification of cost effectiveness of energy systems without CCT and with assessed CCT;
- Selection of the most suitable technology via a combined analysis of the environmental and economic dimensions.

The paper is structured as follows: Section 2.1 defines the LCA methodology, Section 2.2 defines the economical assessment method, Section 3.1 provides a technical description of case study 1, Section 3.2 provides a technical description of case study 2, Section 3.3 defines the systems boundaries, Section 3.4 describes a life cycle inventory, Section 3.5 defines the cost effectiveness parameters, Section 4.1 presents the results of the life cycle impact assessment. Section 4.2 presents the Pareto analysis, Section 4.3 presents the results for cost effectiveness comparison, and Sections 5 and 6 provide the discussion and conclusions.

2. Materials and Methods

2.1. Life Cycle Assessment

The environmental assessment of this study was made using a life cycle assessment (LCA). Life cycle is comprised by the materials extraction, whole supply chain of materials and energies, production process of the specific product, utilization, and end of life. Therefore, the LCA method is considered as a "cradle to grave" analysis. The LCA method consists of the following four steps, defined by the ISO 14,040 standards [11]: goal and scope definition, inventory analysis, life cycle impact assessment, and interpretation.

2.1.1. Goal and Scope Definition

This step outlines the depth of the study by defining the assessed scenarios, their system boundaries, and functional unit. This study aims to assess and compare the environmental impacts of CO_2 capture of two scenarios:

(1)   Scenario 1 defines IGCC pre-combustion CO_2 capture via CaO looping;
(2)   Scenario 2 defines post-combustion CO_2 capture via activated carbon adsorption.

The functional unit (FU) definition provides a reference to which all inputs and outputs of the system are calculated. Thus, FU allows the comparison between the different systems. In this study, FU is defined as 1 kWh of produced electricity, the usual definition of FU for energy systems. The system boundaries describe the processes and modules of the assessed systems included (and excluded) in the environmental model.

2.1.2. Life Cycle Inventory (LCI)

This phase aims to model the system via data collection. Data collection must follow the system boundaries and FU definition. Both scenarios are based on real case studies data from the operational power units. Data was collected within the time frame of one year. For the CCT units, data was collected from the technical project reports [12,13].

2.1.3. Life Cycle Impact Assessment (LCIA)

LCIA comprises classification, characterization, normalization, and weighting process, where the energy and mass flows of the previous step are transformed into environmental impacts. These environmental impacts are calculated according to the selected impact method. Each substance of the assessed systems is multiplied by the characterization factors that determine the potential contribution to the specific environmental impact. An optional step for LCIA is normalization. Normalization enables the comparison between different impact categories. Normalization uses the dataset of the reference indicators of environmental impacts for the European region or worldwide. Thus, the results are values that show the contribution to the sum of European (or world) impacts in the specific environmental category [14].

In this study, the chosen LCIA method is the ReCiPe v.1.08 method of GaBi software at the midpoint level. Based on Carvalho et al. [2], the ReCiPe method is the LCIA, which is intended and tailored for the comprehensive environmental process impact assessment. Also, the ReCiPe method is highly recommended by the EU commission [15]. The characterized midpoint environmental indicators are ozone depletion (OD), human toxicity (HT), ionizing radiation (IR), photochemical oxidant formation (POF), particulate matter formation (PMF), terrestrial acidification (TA), climate change (CC), terrestrial ecotoxicity (TET), agricultural land occupation (ALO), urban land occupation (ULO), natural land transformation (NLT), marine ecotoxicity (MET), marine eutrophication (ME), fresh water eutrophication (FE), fresh water ecotoxicity (FET), fossil depletion (FD), metal depletion (MD), and water depletion (WD) [16].

2.1.4. Interpretation

In this phase, the results are further processed and discussed. The interpretation identifies the significant environmental problems and suggests the optimization of the process toward lowering the impacts. Moreover, it describes the hotspots from assessed processes and indicates significant impact categories. The identification of the significant environmental problems can be done through a Pareto analysis, using the statistical Pareto rule (80/20 rule) [17]. It determines that 20% of all impact categories contributes to 80% of the total environmental impact [2]. Data for Pareto analysis is normalized.

To summarize, the LCA method used in this study was performed by the ReCiPe method. The characterization and normalization were done according to the midpoint level of ReCiPe v 1.08. Moreover, additional Pareto analysis were done to specify key environmental impacts, and further analysis of the concrete processes was made to identify the influence on the environmental impacts.

2.2. Economical Feasibility of the CCT Integration

The economical assessment is the feasibility evaluation required for the comprehensive comparison of the considered technologies. The aim of this economical study is to analyze and compare the cost effectiveness of the reference energy units (REUs) without CCT systems and REU with integration of CCT. The key parameters to compare cost effectiveness are the costs of 1 ton CO_2 separated (for REUs with CCT) and emitted (for REUs without CCT). To perform a cohesive economical assessment, this assessment is based on the technical report [18], which combines several economical international standards and methods tailored for the considered CCT. The economical assessment for this study is based on basic parameters such as

- Capital expenditures (CAPEX);
- Operational expenditures (OPEX);
- Cost of electricity (COE);
- Capture cost (CCo);
- Avoided cost (AvCo).

CAPEX represents the capital expenditures required for the construction of the CCT as a completely new technology or as a retrofit of current technology. This study considers the construction of new technologies from the "greenfield" in both case studies. The costs of the required systems are taken from the market offers of the suppliers of the technological subsystems [19].

OPEX represents the sum of all the operational costs in the first year of the system's operation. The operational costs for energy systems include fuel costs, costs for each media (water, sorbents, desulphurization media, etc.), waste management costs, personal costs, and maintenance costs. Data for the OPEX calculations was taken from the literature sources and real operational data taken by experts from the Czech energy group UJV [18,19].

COE reflects the cost of the electricity produced by the energy source. This criterion shows a simplified view on the economic efficiency of the considered energy source. Thus, if the COE of the assessed energy source is lower than the actual market price of the electricity (in the specific year of operation), the energy source is economically effective (and would therefore generate profit).

CCo represents all the costs required for the separation and capture of 1 ton of CO_2. For REUs without CCT, CCo is determined by the price of CO_2 allowance. The correlation between CCo of carbon capture systems and the market trend of CO_2 allowance determines the cost effectiveness of the carbon capture systems. Moreover, the cost effectiveness of the carbon capture systems shows potential competitiveness in the CO_2 trading market.

AvCo represents the equivalent of CO_2 emissions allowance costs. AvCo defines the costs of 1 tone CO_2 emitted to air.

3. Case Study Definition—Technological Possibilities for CO_2 Emissions Reduction in the Czech Republic

Two innovative technologies via IGCC integrated CaO looping and CO_2 adsorption on active carbon will be explored in the context of Czech coal power units. The adsorption process for the Czech conditions was described in detail in the recent publication of Zakuciová et al. [4].

3.1. Case Study 1—IGCC Power Plant with CaO Looping

The case study is represented by the steam-gas cycle of gross power output of 392 MWe connected to the pre-combustion technology of the integrated gasification combined cycle (IGCC) with integration of the carbonate loop. IGCC uses the high pressure gasifier to produce pressurized gas (synthesis gas) from the carbon-based fuels [12]. The principle of the system is based on steam-gas cycle with hydrogen combustion and with integrated gasification of lignite with CO_2 capture from the gas before combustion. The IGCC system ensures the removal of impurities such as sulfur dioxides and particulates from the syngas before the actual carbonate looping. The case study represents a specific technology, where the elimination of the acid impurities is based on high temperature desulphurization by adsorbent of $CaO/CaCO_3$.

The main advantage of the calcium looping system is the high degree of CO_2 removal (up to 95%) and the process of simultaneous desulfurization [20].

To understand and define the system's boundaries while comparing scenarios, it is important to describe both scenarios from technical point of view.

The IGCC process can be divided into the following technological segments (Figure 1):

1. Management and treatment of the fuel;
2. Oxygen production—cryogenic separation;
3. Gasification process—shell gasification technology;
4. Purification of the synthetized gas (high temperature desulphurization and ceramic filter for particulates separation), water gas shift reaction, and CO_2 separation by carbonate looping;
5. Energetic utilization of synthetized gas with high H_2 content (steam–gas cycle).

3.1.1. Fuel Management and Treatment

The management of the fuel comprises lignite mining, transportation, storage, crushing (max. 40 mm), drying and grinding (max. 200 μm). Lignite is expected to be mined from the ČSA (Karviná region) quarry due to specific parameters (low concentration of ash). The lignite is then transported by railways to the storage located next to the power unit, ground, and dried to 200 μm with a maximum level of moisture (11%). The process of lignite drying uses the energy from the steam produced in the steam-gas cycle. For drying we expect use WTA (waste heat utilization) technology.

3.1.2. Oxygen Production

Oxygen will be produced in the separated oxygen unit. The recommended process of the oxygen separation from the air is the cryogenic separation, a well-known and viable process. The electricity used for the cryogenic separation will be generated from the steam–gas cycle. The main outputs from the cryogenic separation are oxygen with 95% purity and nitrogen with purity of 98.7%. The nitrogen is then mixed with hydrogen as a fuel to the steam–gas cycle.

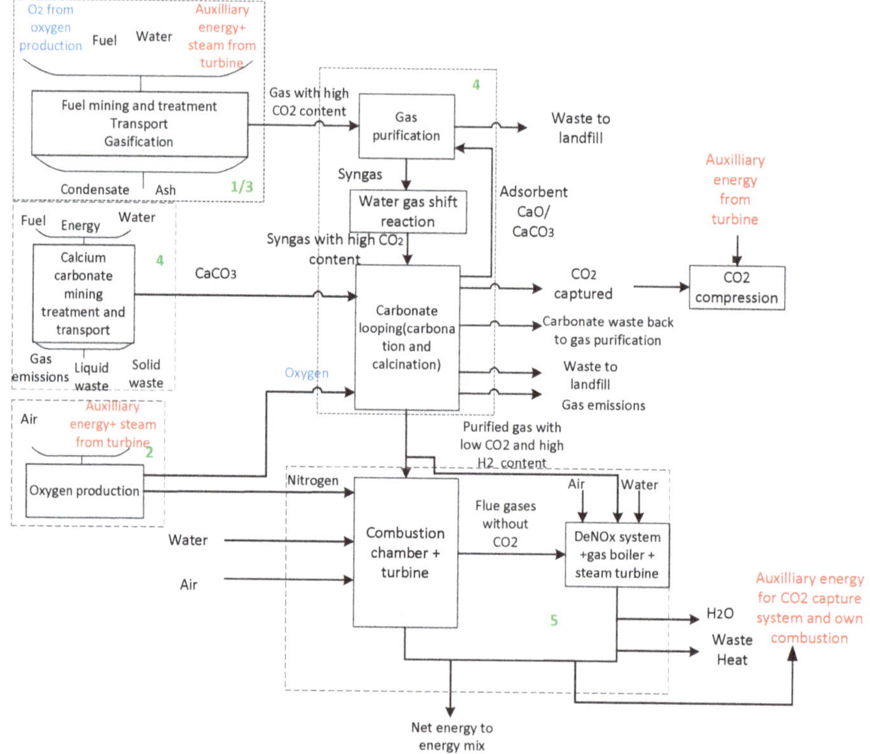

Figure 1. System boundaries for Scenario 1 (marked red are energies returning back to the system; dashed lines with numbers represent technological segments).

3.1.3. Gasification Process

The gasification process of lignite is operated in a Shell generator (considered a modern and verified technology for this process). This type of generator ensures the lowest content of organic compounds that can create problems in the further purification process. In the generator, the oxygen reacts with the lignite (chemical reactions (1) and (2) [21]) in an exothermic reaction, creating a temperature around 1500 °C. In this temperature, the ash from the fuel is transformed into liquid slag. Gas coming out of the generator is cooled down by the injection of water to the temperature of 900 °C.

$$C + O_2 \rightarrow CO_2 \ \Delta H = -394 \ kJ/mol \tag{1}$$

$$C + 0.5 \ O_2 \rightarrow CO \ \Delta H = -111 \ kJ/mol \tag{2}$$

3.1.4. High-Temperature Purification Process, Water Gas Shift Reaction, and Carbonate Looping

The high-temperature purification process includes high-temperature desulphurization at temperatures between 800–900 °C. Desulphurization is done via adsorption of all the acidic impurities (e.g., H_2S) on sorbent $CaO/CaCO_3$ (reaction (3)) [22] that comes from the carbonator. The waste product after the purification process is a mix of $CaCO_3 + CaSO_4$, which is transported as waste to a landfill.

$$CaO + H_2S \rightarrow CaS + H_2O \tag{3}$$

The output from the purification process is purified gas. The gas is then transported into the water–gas shift reactors where the shift reaction is achieved. Said reaction (4) [21] converts CO into CO_2 by steam. The purified gas after the shift reaction contains a higher rate of CO_2.

$$CO + H_2O \rightarrow CO_2 + H_2 \; \Delta H = -41 \text{ kJ/mol} \tag{4}$$

After the shift reaction, the gas is transported into the carbonate loop system, where the CO_2 is separated. At first, the gas enters the carbonator. In the carbonator the exothermic reaction of CaO with CO_2 takes place (reaction (5)) [21].

$$CaO \text{ (s)} + CO_2 \text{ (g)} \rightarrow CaCO_3 \text{ (s)} \; \Delta H = -178.2 \text{ kJ/mol} \tag{5}$$

The temperature in the carbonator should not exceed 800 °C. The gas after the carbonation process proceeds into the combustion chamber with turbine.

The produced $CaCO_3$ from the carbonator is transported into the calcinator that works in the oxyfuel regime. The temperature in the calcinator increases to 950 °C and the $CaCO_3$ is decomposed back into CaO and CO_2 (reaction (6)) [21]. CaO returns into the carbonator to be used as sorbent. Moreover, a fresh batch (2.5 t/h) of $CaCO_3$ is periodically (once in 20 min) added into the calcinator.

$$CaCO_3 \text{ (s)} \rightarrow CaO\text{(s)} + CO_2 \text{ (g)} \; \Delta H = -178.2 \text{ kJ/mol} \tag{6}$$

The emissions from the calcination process (mainly CO_2) are cooled and compressed. The liquefied CO_2 is separated. The CO_2 compression requires auxiliary energy, provided from the steam–gas cycle.

3.1.5. Energetic Utilization of Synthetized Gas

Gas with high H_2 content (after the purification, shift and CO_2 capture) will be mixed with nitrogen (waste product of oxygen production). The mixture of the synthetized gas with nitrogen ensures the high energy efficiency of the whole system. Thus, the net calorific value of the synthetized gas must not be lower than 12.8 MJ/kg (6.9 MJ/m^3).

The unique principle of the case study was designed and tailored for the conditions of Czech operational power unit. The whole concept was designed by the biggest energy research company in the Czech Republic (UJV group) for the national project [12]. The advantage of this system is not requiring such a high external energy input. On the other hand, IGCC-CaL system lowers the power generation efficiency to 25.3%.

3.2. Case Study 2—Activated Carbon Adsorption

Case study 2 considers the thermal power unit with the gross output of 250 MWe connected to post-combustion carbon capture technology based on adsorption. The adsorption unit was designed as a pilot facility to capture CO_2 from operational flue gases. It is based on a rotative adsorber of continuous operation. The rotative adsorber operates in three phases of adsorption, desorption, and cooling. In the operation (even with minimum concentration of pollutants in the flue gases entering adsorber), the sorbent will degrade, and it is necessary to periodically it with 23 kg/h of the activated carbon pellets. The source for active carbon production is hard coal. Hard coal is further processed in two steps of (1) carbonization of the raw hard coal without presence of oxygen and (2) activation of the carbonized product by water vapor. The whole process chain can be divided into technological segments as follows:

1. Preparation of fuel for the power unit;
2. Preparation of solvent for flue gas purification and cooling;
3. Preparation of activated carbon for CO_2 capture;
4. Operational part of the power unit;

5. CO_2 capture and compression.

The technological details of the whole technology are described in Zakuciová et al. [4] and the technological segments are shown in Figure 2. The advantage of this process is the continuous operation and higher power generation efficiency of 33.73%. However, the activated carbon production requires a process of activation and carbonization that consumes more raw material (hard coal) and energy.

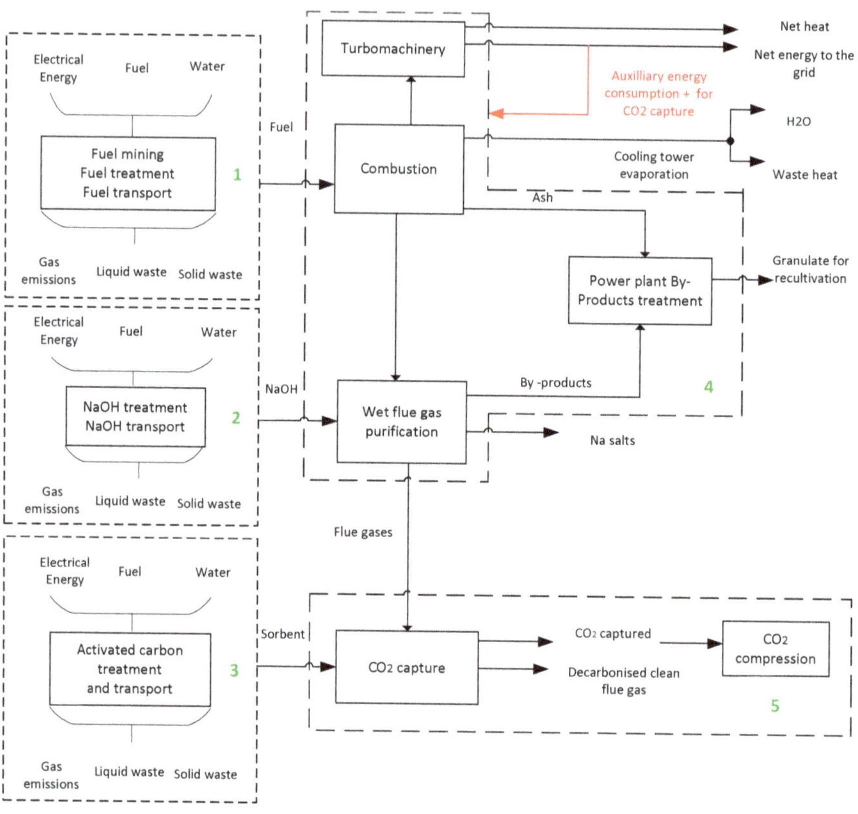

Figure 2. System boundaries for Scenario 2 (red marked is the energy returning back to the system; dashed lines with numbers represent technological segments).

3.3. LCA Study—System Boundaries Definition

The system boundaries for both scenarios are based on the technological description of each system (Figures 1 and 2). Scenario 1 (case study 1) includes all the described technological segments of fuel management, carbonate loop with carbonate production chain, gasification and purification processes, combustion of the syngas, steam gas cycle with electricity production, and CO_2 capture.

Scenario 2 (case study 2) was proposed in a previous study [4]. The system boundaries for the adsorption process includes the fuel supply chain for the power unit, sorbent supply chain for adsorption process, and the NaOH treatment process. Moreover, the boundaries consider the operational part of the power unit including energy production, fuel combustion, flue gas treatment, CO_2 adsorption process, and treatment of the waste products.

Both scenarios are including CO_2 compression (CO_2 compressed to 10–11 MPa). The further CO_2 transport and storage are not included in the system boundaries due to lack of current scenarios for the storage of the captured CO_2 in Czech conditions. The assumption is to sell the captured CO_2 as a viable product for greenhouse farming or algae-based CO_2 conversion.

3.4. Life Cycle Inventory

The inventory data for both scenarios are based on the Czech operational power units in the conditions of the Czech national energy mix. The actual operational data was taken from the reference power unit of 250 MWe. In the case of post-combustion by activated carbon adsorption, data was calculated to comply with the dimensions of the 250 MWe power unit. In the case of IGCC-CaL, the whole system was designed and simulated for dimensions comparable with the 250 MWe power unit.

Inventory data for scenario 1 (Table 2) was collected from data calculated by experts based on operational parameters of the power unit. The data is the subject of the national research project [12].

Table 2. Basic inventory data for scenario 1.

Processes	Inputs	Amount	Units	Outputs	Amount	Units
Fuel treatment and gasification	Brown coal (dried)	175.3	t/h	Ash	39.19	t/h
	O_2	102.88	t/h	Syngas	425.73	t/h
	Water	150.4	t/h			
Gas purification	CaO	7.65	t/h	$CaCO_3 + CaSO_4$ (waste)	11.75	t/h
Oxgen production	Air	864,767	m^3/h	O_2	181,169	m^3/h
	Electricity	90.2	kW	N_2	675,556	m^3/h
Carbonate looping	O_2	95.52	t/h	CO_2 captured	290.97	t/h
	$CaCO_3$ total batch	130	t	CO with flue gases into air	13.16	t/h
	Syngas	234.07	t/h			
	$CaCO_3$ fresh sorbent input	25	t/h			
Steam-gas cycle	Syngas (energy efficiency)	12.8	MJ/kg	Nominal power output (without CCT)	381.71	MWe

The composition of syngas [12] is described in the following table (Table 3).

Table 3. Syngas composition after gasification and carbon dioxide removal.

Syngas after Gasification Process	t/h	Syngas after Shift Reaction and CO_2 Removal	t/h
CO	202.58	CO	2.25
CO_2	34.65	CO_2	9.61
H_2	9.94	H_2	13.43
N_2	18.84	N_2	10.44
H_2O	150.4	Ar	2.86
Others	9.32	Sum	38.59

Inventory data for the scenario 2 is described in Table 4. Inventory data for PCC-A was (like in case study 1) optimized by calculations based on the operational parameters of the power unit. Data for operational power unit without CCT was obtained from the actual operation of 250 MWe unit [13].

The following table (Table 5) shows the differences between both scenarios in terms of energy consumption for both systems with implemented carbon capture systems, percentage of captured CO_2 and differences in the thermal efficiency before and after carbon capture implementation.

Table 4. Basic inventory data for scenario 2.

Processes	Inputs	Amount	Units	Outputs	Amount	Units
Power unit	Brown coal (dried)	214	t/h	Nominal power output (without CCT)	226	MWe
	Water	9258.63	t/h	Condensate and wastewater	145	t/h
Gas purification	NaOH	7.65	t/h	Reactive products	0.526	t/h
Activated carbon production	Hard coal	190	t	Tar	76	t
	Energy for activation	1132	MJ			
CO_2 adsorption	Activated coal total batch	760	t	Flue gases	685,955	m^3/h
	Flue gases	766,045	m^3/h	CO_2 captured	158	t/h
	CO_2	211	t/h			

Table 5. Differences in significant technological parameters of both scenarios.

	Scenario 1	Scenario 2
Power consumption for CO_2 capture and compression	119.31 MWe	28 MWe
CO_2 capture ratio	95%	75%
Thermal efficiency	**Without CO_2 capture** 37.80%	38.40%
	With CO_2 capture system 25.30%	33.73%
Specific power consumption (MWe/t CO_2 captured)	0.9	1.3
Nominal power output	262.4 MWe	198 MWe

Moreover, these additional assumptions were taken into consideration for both scenarios:

- Energy required for activated carbon production and calcium carbonate production is based on natural gas;
- Carbonate waste and reactive products are considered as a waste for landfill;
- CO_2 captured is considered as a valuable product for further utilization;
- Transport distances by diesel from mining quarries to power unit are modelled as average distance of 500 km;
- Wastewater is considered for the further treatment in the wastewater treatment plant data taken from the database of EU standard of the Thinkstep dataset;
- Process of oxygen production via cryogenic separation is based on database process from the EU28 Thinkstep database;
- Water for the power units is considered as processed water (demineralized, deionized);
- Production chain of materials such as limestone, NaOH, and oxygen are taken from the EU standard of the Thinkstep dataset;
- Specific regional production a mix of hard coal and lignite is taken from the Czech and Slovak Thinsktep dataset.

3.5. Economical Assessment-Cost Effectiveness Parameters Definition

As stated in Section 2.2, CAPEX, OPEX, and COE are the basic parameters for the cost effectiveness comparison. It is important to mention that CAPEX and OPEX are different for both REUs that are prepared for IGCC-CaL and adsorption integration. Apart from the difference in the technological segments, which influence CAPEX data [18,19], also OPEX variables such as fuel costs, solvent/adsorbent cost and final cost of the produced electricity vary for each REU system (Table 6). However, the operational time for both systems is assumed to be equal (7400 h/year). CAPEX and OPEX for the REUs and REUs with CCT systems are summarized in the Table 7.

Table 6. Operational expenditures (OPEX) variables for reference energy units (REUs) of both case studies [18,19].

Costs	REU (IGCC-381.71 MWe)	REU (Sub-Critical Coal Power Unit-226 MWe)
Fuel cost (€/t)	35.76	23.07
Cost of solvent/adsorbent (€/kg)	0.34	0.76
Market price of electricity (€MWh)	45	45
Fixed operational costs (€/year)	2,692,308	1,507,692

Table 7. Capital expenditures (CAPEX) and OPEX for REUs without and with carbon capture technology (CCT) systems [18,19].

Technological Segments	REU (IGCC)	REU (Sub-Critical Coal Power Unit)	REU + CCT (IGCC-CaL)	REU + CCT (PCC-A)
	381.71 MWe	226 MWe	262.4 MWe	198 MWe
CAPEX (million Euros)	1025.9	716.9	1264.1	1097
OPEX (million Euros)	114.1	121	140.3	123.05

COE is based on the following Equation (7):

$$COE = \frac{I_0(t=1) + O_{fix} + O_{var}}{P_e} \tag{7}$$

where

I_0 is the modified ratio of capital expenditures that refer to 1 year of the operation (€/year);

O_{fix} are fixed operational costs (e.g., costs for maintenance and repairs) (€/year);

O_{var} are variable operational costs (e.g., fuel costs) (€/year);

P_e amount of produced and delivered electric energy to the net in the first year of operation (MWh/year).

The cost effectiveness is based on Capture cost (*CCo*) and is calculated as follows Equation (8):

$$CCo = \frac{COE_{with\ CCT} - COE_{without\ CCT}}{amount\ of\ separated\ CO_2} \tag{8}$$

For REU, the cost effectiveness is based on the price of the CO_2 allowance. This study is taking into account the market trend of the price of CO_2 allowances for the time frame of 2015–2050 [23–26]. The parameter of avoided cost of emitted CO_2 is expressed as follows Equation (9):

$$AvCo = \frac{COE_{with\ CCT} - COE_{without\ CCT}}{emissions\ CO_{2without\ CCT} - emissions\ CO_{2with\ CCT}} \tag{9}$$

4. Results

This section presents the results according to the methodology for LCIA and economic methodology. Further analysis of the results is discussed in detail in Section 5.

4.1. Life Cycle Impact Assessment

The results for both scenarios (Table 8) are divided into three groups of values:

1. Characterization values of the environmental category;
2. Normalized values according to ReCiPe 1.08 Midpoint normalization for the European region (units for all the impact categories are points);
3. Relative contribution of each environmental category to the sum of all impact categories based on normalized values.

Table 8. Results for both scenarios (EIC—environmental impact categories, CHV—characterization values, NV—normalization values, RC—relative contribution); ozone depletion (OD), human toxicity (HT), ionizing radiation (IR), photochemical oxidant formation (POF), particulate matter formation (PMF), terrestrial acidification (TA), climate change (CC), terrestrial ecotoxicity (TET), agricultural land occupation (ALO), urban land occupation (ULO), natural land transformation (NLT), marine ecotoxicity (MET), marine eutrophication (ME), fresh water eutrophication (FE), fresh water ecotoxicity (FET), fossil depletion (FD), metal depletion (MD), water depletion (WD).

	Scenario 1			Scenario 2		
EIC	CHV	NV	RC in %	CHV	NV	RC in %
ALO	1.54×10^{-3} (m²a)	1.18×10^{-6}	0.05	9.70×10^{-3} (m²a)	2.15×10^{-6}	0.05
CC	1.54×10^{-1} (kg CO_2 eq.)	1.38×10^{-5}	0.6	5.72×10^{-1} (kg CO_2 eq.)	5.10×10^{-5}	1.21
FD	1.96×10^{-1} (kg oil eq.)	1.26×10^{-4}	5.46	8.58×10^{-1} (kg oil eq.)	5.50×10^{-4}	13.03
FET	2.62×10^{-5} (kg 1.4 DB eq.)	2.40×10^{-6}	0.1	4.74×10^{-4} (kg 1.4 DB eq.)	4.35×10^{-5}	1.03
FE	2.17×10^{-7} (kg P eq.)	5.24×10^{-7}	0.02	4.19×10^{-6} (kg P eq.)	1.01×10^{-5}	0.24
HT	2.01×10^{-3} (kg 1.4-DB eq.)	3.40×10^{-6}	0.15	4.38×10^{-2} (kg 1.4-DB eq.)	7.40×10^{-5}	1.75
IR	2.33×10^{-2} (U235 eq.)	3.72×10^{-6}	0.16	1.62×10^{-2} (U235 eq.)	2.59×10^{-6}	0.06
MET	1.07×10^{-5} (kg 1.4-DB eq.)	1.26×10^{-6}	0.05	8.02×10^{-5} (kg 1.4-DB eq.)	9.44×10^{-6}	0.22
ME	1.56×10^{-3} (kg N eq.)	1.55×10^{-4}	6.71	2.43×10^{-3} (kg N eq.)	2.41×10^{-4}	5.71
MD	4.48×10^{-4} (kg Fe eq.)	6.28×10^{-7}	0.03	7.07×10^{-3} (kg Fe eq.)	9.91×10^{-6}	0.23
NLT	1.03×10^{-6} (m²)	6.38×10^{-6}	0.28	1.89×10^{-5} (m²)	1.17×10^{-4}	2.77
OD	2.31×10^{-13} (kg CFC–11 eq.)	1.05×10^{-11}	0	2.15×10^{-13} (kg CFC-11 eq.)	9.76×10^{-12}	0
PMF	8.80×10^{-3} (kg PM10 eq.)	5.91×10^{-4}	25.59	1.37×10^{-2} (kg PM10 eq.)	9.19×10^{-4}	21.77
POF	3.99×10^{-2} (kg NMVOC eq.)	7.52×10^{-4}	32.56	6.20×10^{-2} (kg NMVOC eq.)	1.17×10^{-3}	27.72
TA	2.24×10^{-2} (kg SO_2 eq.)	6.52×10^{-4}	28.23	3.49×10^{-2} (kg SO_2 eq.)	1.02×10^{-3}	24.16
TET	1.13×10^{-6} (kg 1.4-DB eq.)	1.38×10^{-7}	0.01	9.20×10^{-6} (kg 1.4-DB eq.)	1.12×10^{-6}	0.03
ULO	1.05×10^{-4} (m²a)	2.57×10^{-7}	0.01	1.91×10^{-4} (m²a)	4.68×10^{-7}	0.01
WD	2.40×10^{-1} (m³)	0.00	0	2.99×10^{-1} (m³)	0.00	0
SUM	-	2.31×10^{-3}	100	-	4.22×10^{-3}	100

The characterization values show results in absolute values (first column for scenario 1 and 2), comparable only within one impact category.

The normalized results (second column for scenario 1 and 2) allow us to compare the severity of environmental impact categories among them, as all the impact categories are calculated in common units (points).

The relative contribution (third column for scenario 1 and 2) helps to identify a contribution of each environmental category in a certain ratio (%) to the sum of normalized values of all impact categories (100%). Relative contribution of each environmental category computes according to Equation (10).

$$Cont_{ic} = \frac{NV_{ic}}{SUM_{NV}} \times 100 \tag{10}$$

where

$Cont_{ic}$ is the relative contribution of each environmental impact category to the sum of environmental impacts;

NV_{ic} is the normalized value of the specific impact category;

SUM_{NV} is the total sum of the normalized values of all impact categories.

The sum of all normalized values indicates that scenario 1 has lower environmental impacts than scenario 2 and, therefore, better environmental performance. From Table 8, it is possible to verify that

POF, TA, and PMF show the highest relative contributions regarding the total environmental impact of the systems.

In scenario 1, marine eutrophication and fossils depletion have also (6.71% and 5.46%) notable relative contribution. All other categories have relative contribution below 1%.

Categories of CC and FD are also important to analyze. The assessed systems are dealing with CO_2 capture; thus, the category of CC is directly influenced. Also, the brown coal mining and treatment are key processes which influence the category of FD.

4.2. Pareto Analyses and Comparison of the Processes among Scenarios

As previously stated, Pareto analysis helps to define the environmental categories of the highest significance to the total of environmental impacts. Figures 3 and 4 show the key impact categories for each scenario. Both scenario 1 and scenario 2 identify the most significant environmental categories of POF, TA, and PMF. Both figures show just the visible values on the plot. The remaining impact categories have a relative contribution below 1%.

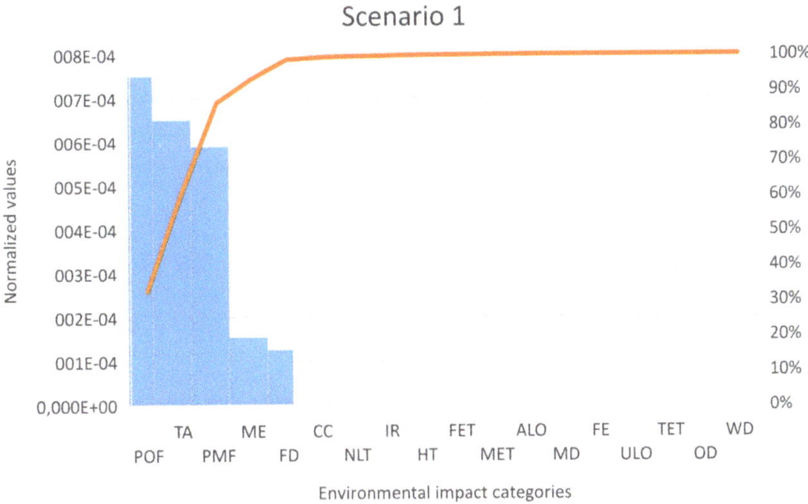

Figure 3. Pareto analysis for scenario 1 (only visible values).

In the next step, it is important to define the hotspots in the processes (most impactful processes) for both scenarios. Therefore, further analysis of the potential contribution of the processes for the most critical impact categories was performed. In this analysis, the categories cannot be compared between each other. The analysis is based on the characterization values, and therefore, it is focused on one impact category at the time, influenced in different ratios (%) by different processes. The results for the processes contribution are summarized in Table 9.

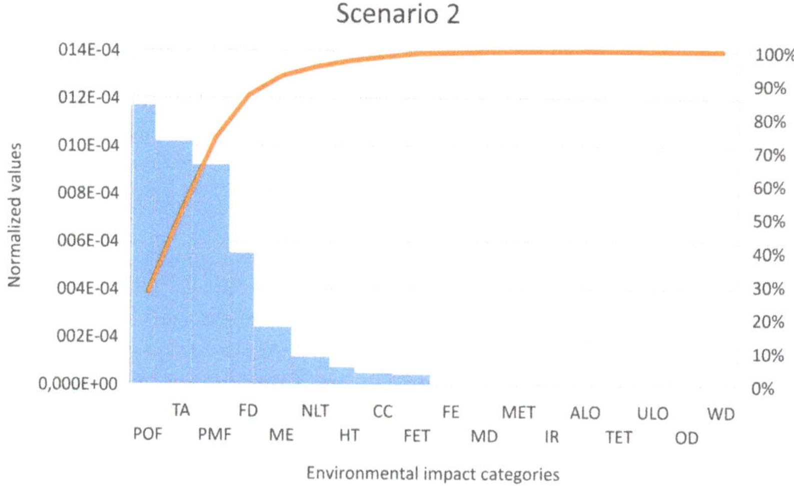

Figure 4. Pareto analysis for scenario 2 (only visible values).

Table 9. Relative contribution of the processes to significant environmental categories.

	Scenario 1				Scenario 2			
	Lignite Mix	Oxygen Production by Cryogenic Separation	Limestone Mining and Treatment	Carbonate Looping	Lignite Mix	Hard Coal Mix for Activated Carbon	Process Water for Cooling and Purification	Adsorption	Thermal Energy from Natural Gas Mix
POF	0.10%	0.10%	0.008%	99.60%	0.18%	1%	0.20%	96%	1.80%
TA	0.10%	0.47%	0.01%	99.40%	0.14%	2%	0.36%	95%	2.10%
PMF	0.10%	0.36%	0.05%	99.40%	0.13%	2%	0.30%	95%	2%
ME	0.12%	0.66%	0.15%	98%	0.10%	1.30%	1.02%	95%	1.16%
FD	86.70%	9.40%	1.66%	-	19.10%	40.70%	1.70%	-	40%
NLT	4.50%	80.90%	10.30%	-	0.60%	94%	5.00%	-	-
CC	66.14%	28.40%	4.44%	0.02%	7.65%	16.50%	3.38%	0.05%	72%

In the category of climate change, the flow of captured CO_2 is referred to as environmental credit. However, in both cases, the CO_2 emissions and clean flue gases are still released to the air and therefore contribute to the environmental impacts. Some of the processes do not have any relative impact to the environmental categories.

4.3. Cost Effectiveness Comparison

According to the Equations (7)–(9), COE, CCo, and AvCo for both IGCC-CaL and PCC-A systems are summarized in the following table (Table 10).

Table 10. Results of cost of electricity (COE), CO_2 capture cost, and avoided CO_2 cost.

Parameters	IGCC-CaL	PCC-A
COE (€/MWh)	123.1	90.24
CO_2 capture cost (€/t CO_2)	57.1	37.48
Avoided cost (€/t CO_2)	105	34.06

The cost effectiveness is shown in Figure 3. The figure shows the correlation between the carbon price on the market (black lines) from 2015 up to 2050 [23–26] and CO_2 avoided costs of both CCT systems connected to REU. The comparative economic criteria were defined/re-calculated with respect to the inputs and variables based on 2018. Two primary cases were analyzed. The first case (red and green dotted lines) describes CCT utilization as a key economic unit in carbon capture utilization (CCU) scheme with the possibility of using CO_2 within the enhanced oil recovery, fuel production, etc. The second case (red and green dashed lines) reflected the CCT as a fundamental unit together with transportation and storage in the carbon capture and storage (CCS) scheme. The Czech Republic considers CO_2 transport by pipelines into salt aquifers in the Zatec Basic (North-West Bohemia, the Czech Republic) [27,28].

The carbon price curve demonstrates the possible development of the market of carbon price in 2015–2050. The proposed estimation in Figure 5 was defined as a combination of the real average annual data from the market (black line) and an estimate based on CAKE/KOBiZE forecast (dashed black line). Moreover, the initial CAKE/KOBiZE forecast [25] is also displayed in Figure 5 (dotted black line). This forecast was evaluated based on the Paris Agreement for the Central Europe power sector (more precisely Poland).

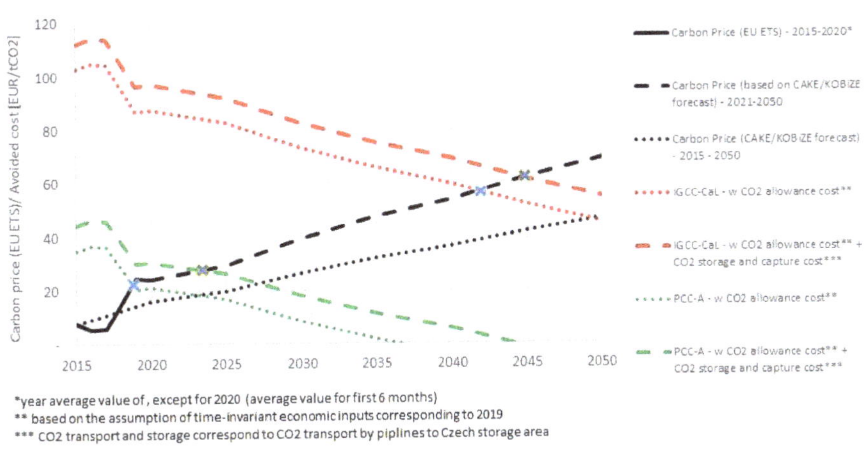

Figure 5. The cost-effectiveness of selected CCT integration into power plants.

5. Discussion

The discussion part follows the sections of the results. At first, the environmental assessment with processes analysis will be discussed. In the second part, the economical assessment will be analyzed. In the last part, the combination of environmental and economic results will be concluded.

5.1. Environmental Assessment

The results of the characterization, normalization, and relative contribution of the environmental impacts are shown in Table 8. The absolute values of the characterization process enable the comparison of the same environmental impact category among scenarios. At first glance, scenario 1 has a lower characterization values in comparison with scenario 2 in almost all environmental categories except in the category of ionizing radiation. Ionizing radiation in scenario 1 is influenced by the process of oxygen production via cryogenic separation. This process is the database process and does not reflect the local impact. Moreover, the radiation impact has an insignificant contribution in comparison with other impacts (sc.1 0.16%; sc.2 0.06%).

When we aggregate the environmental impact categories after normalization (see Table 8 sum of normalized values), the scenario 2 has a higher global environmental impact (0.00231) than scenario 1 (0.0042). However, it is important to stress out that the case studies considered within this manuscript are different in several parameters such as (i) different scale of REU, (ii) site specific case studies (iii) different power generation technology. Therefore, the environmental performance for scenario 2 might be improved regarding LCA results if the REU technology would have the same technological basis in both cases.

However, it is important to analyze the contribution of the individual impact categories in the total environmental impact. To analyze the highest significance of the impact categories of each scenario, the pareto analysis was chosen as a decision-making tool (Figures 3 and 4). Both figures are confirming the relative contribution results that impacts category of POF, TA and PMF have the highest contribution to the sum of all normalized values. Nevertheless, in comparison to scenario 2, scenario 1 has in the category of ME a slightly higher contribution of 1%, and in the category of FD, it has a lower contribution of 7.57%. Another difference is seen in the category of NLT, where scenario 2 exceeds scenario 1 by 2.5%. As both scenarios have CO_2 capture as primary function, the environmental impact category of climate change (CC) (which measures the contribution of CO_2 and other compounds to the global warming) has a lower contribution to the total environmental impact. This was expected since both technologies capture the CO_2.

The impact categories are influenced by the environmental impact of the different processes in the life cycle. Table 9 shows that the significant impact categories taken from the pareto analysis are influenced by specific processes in both scenarios. According to Table 9, the categories of POF, TA, PMF, and ME are influenced in 95% to 99.6% by the emissions of CO_2 capture process (carbonate looping, adsorption) in both scenarios (Tables 2 and 4). However, the characterization values, for instance in category POF, are very small (scenario 1—0.039 kg NMVOC eq./1 kWh; scenario 2—0.062 kg NMVOC eq./1 kWh).

In the category of FD, the production chain of the lignite from Slovak lignite mix (which has a similar thermal efficiency as Czech lignite of 11 MJ/kg [12]) results as the most significant process in both scenarios. In scenario 1, the lignite mix for the power unit contributes almost 86.40% to FD (Table 9). In scenario 2, the hard coal mix for active carbon production increases the FD contribution by 40.70%, whereas lignite mix for power unit contributes 19.10%. Moreover, in scenario 2, the category of FD is influenced 40% by the utilization of thermal energy from natural gas mix for activated carbon production.

An interesting result is shown for the category of NLT. In scenario 1, major land transformation would be impacted by the construction of an air separation unit for oxygen production. In scenario 2, the hard coal production chain with all the mining process necessary for active carbon production turns out to be the process with the highest impact to the natural land transformation. Moreover, in comparison with scenario 1, hard coal would need to be mined and transported to the power unit, which creates an additional environmental burden. In scenario 1, the air separation unit would need to be built right in the local area of the power unit.

The category of CC is mainly influenced by the ratio of captured CO_2. It is obvious that adsorption process would require higher amount of active carbon to be able to capture 95% of CO_2 such as an IGCC-CaL system. That would lead to the increase of the total environmental impact. Also, the thermal energy from the natural gas combustion as a primary energy for activated carbon production, is influencing category of CC in 72%.

The primary goal for the CCT solution under Czech energy conditions was to design and compare post combustion and pre combustion systems for the same REU. However, during the research, problems occurred with the technological requirements (such as quality of the lignite for each REU) of each system. Thus, the input parameters had to be optimized, which led to different scale of REU, different lignite quality and different technological segments. Therefore, the specific case studies considered in this manuscript do not have the same basis for fair comparison. Still, in LCA analysis relating all the environmental impacts to 1 kWhe, the aforementioned differences are still present in particular environmental impacts.

This paper considers the specific case of activated carbon production from hard coal. However, activated carbon can be produced in several options from biomass or other organic waste that would decrease total environmental performance of the process. Also, different adsorption process configurations may lead to different results. This specific case of the PCC-A points out the problem with Na salts production after flue gas purification, which are currently considered as waste material with no other use.

The main advantage of PCC-A against IGCC-CaL is the thermal efficiency of the whole system. For the process of the CO_2 capture and compression, PCC-A requires a consumption of 28 MWe from the power unit, whereas IGCC-CaL requires for the same process 119.31 MWe (Table 5). Thus, PCC-A decreases the thermal efficiency of the power unit by 4.67%, and IGCC-CaL decreases the thermal efficiency of the power unit by 12.5%. The thermal efficiency decrease would require a higher energy production that might also influence the environmental performance of IGCC-CaL system as well. Moreover, the thermal efficiency decline may be significant for the further operational costs increase. On the other hand, the specific energy consumption (MWe/t CO_2) in Table 5 states that PCC-A (1.3 MWe/t CO_2) would require a slightly higher energy demand than the IGCC-CaL process (0.9 MWe/t CO_2).

The following table (Table 11) shows the comparison of this study with other studies of [8–10]. The environmental results of this study for pre-combustion IGCC-CaL system shows a lower kg CO_2 eqv./MWh than in a similar study of Petrescu et al. [8]. The lower impact of climate change of this study (global warming potential) corresponds to the smaller size of the reference energy power plant and higher capture ratio. For eutrophication potential this study is resulting in much lower values that are comparable with the study of Clarens et al. [10]. However, the impacts of the acidification potential in this study are the highest among of all studies. This might be influenced by the production of the used sorbent in the form of $CaCO_3$ + $CaSO_4$ as non-utilized waste product of high-temperature desulphurization. Moreover, the study shows the highest drop in net energy efficiency due to CCT implementation. The reason might be that the IGCC-CaL design of this study does not consider utilizing the heat losses due to lack of commercially viable heat exchangers for such amount of heat. In the case of post-combustion capture, PCC-A has the lowest CO_2 capture rate among all the studies. It corresponds to higher values in climate change in comparison with similar study of Clarens et al. [10]. Also, in this study of PCC-A, specific emissions are the highest, which refers to the low CO_2 capture rate.

5.2. Economical Assessment

The key economic parameter for CCT integration is capture cost. The comparison of capture costs (Table 10) among assessed CCT systems states that the less expensive technology is PCC-A system. The difference is shown in the values of CAPEX, which for IGCC-CaL is higher by €167 million and in OPEX by €17 million annually. The main reason for this difference is that PCC-A does not require high technological adjustments when compared to the IGCC-CaL systems. The IGCC-CaL system requires an initial batch of lignite with higher quality for gasification process, therefore the technological components (such as boiler) would increase the initial CAPEX. Moreover, IGCC-CaL requires the construction of an additional segment with auxiliary systems of air separation unit that also increases the initial investment.

Another parameter—cost of electricity (COE)—is influenced by OPEX. The higher OPEX of the IGCC-CaL system increases the energy cost by €33/MWh in comparison with the PCC-A (Table 7). However, both case CCT studies are showing a higher COE per MWh in comparison with the actual market price of electricity (Table 6). That leads to the conclusion that both systems tailored for the current Czech conditions are currently not economically viable.

Table 11. Comparison of study results with several references; GWP/CC—global warming potential/climate change; TA/AP—terrestrial acidification/acidification potential; FE/EP—reshwater eutrophication/eutrophication potential.

Studies	Reference Power Plant/Type/Net Energy Power Output	Type of CCT/Net Energy Power Output	Net Energy Efficiency without CCT (in %)	Net Energy Efficiency with CCT (in %)	CO_2 Capture Rate in %	CO_2 Specific Emissions for CCS	LCA Method and Software	Environmental Impacts
Zakuciová et al. (2020)	Coal based IGCC with Shell gasifier/ 381.7 MWe	Pre-combustion IGCC-CaO looping/ 262.4 MWe	37.8	25.3	95	36.32 kg/MWh	ReCiPe method v 1.08, Gabi Sotware v. 6	CC 154 kg CO_2 eqv./MWh; TA 22.4 kg SO_2 eqv./MWh; FE 0.0002 kg P eqv./MWh
Zakuciová et al. (2020)	Sub-critical brown coal power unit/ 226 MWe	Post combustion adsorption on activated carbon/198 MWe	38.4	33.73	75	267 kg/MWh	ReCiPe method v 1.08, Gabi Sotware v. 6	CC 572 kg CO_2/MWh; TA 34.9 kg CO_2/MWh; FE 0.0042 kg P eqv./MWh
Petrescu et al. (2017)	IGCC power plant with acid gas removal based on Selexol/ 493.13 MWe	Pre-combustion IGCC-CaO looping/ 607.82 MWe	45.09	36.44	91.56	58.87 kg/MWh	LCA method based on CML 2001, GaBi software v. 6	GWP 917.25 kg CO_2 eqv./MWh, AP 1.47 kg SO_2 eqv./MWh, EP 1461.97 kg PO_4^{3-} eqv./MWh
Petrescu et al. (2017)		Pre-combustion IGCC-FeL based oxygen carriers chemical looping/ 443.07 MWe	45.09	38.76	99.45	3.01 kg/MWh		GWP 338.73 kg CO_2 eqv./MWh, AP 1.75 kg SO_2 eqv./MWh, EP 1949.12 kg PO_4^{3-} kg eqv./MWh
Rolfe et al. (2017)	Supercritical pulverized coal power plant with flue gas desulphurization and selective catalytic reduction/ 588.6 MWe	Post combustion CaO looping/ 924 MWe	39.1	32.1	90	88 g CO_2/kWh	ReCiPe method, SimaPro software v 8.3	Not indicated absolute values; reduction in 72% GWP; 9% increase in resource use
Clarens (2016)	Sub-critical power plant/ 469.9 MWe	Post combustion CaO looping/ 519.4 MWe	36.9	29.6	90	N/A	ReCiPe midpoint method v1.04, SimaPro software	CC 0.26 kg CO_2 eqv./kWh; TA 0.001 kg SO_2 eqv./kWh; FE 0.0051 kg P eqv./kWh

The economical assessment of this study is based on the comparison of the cost effectiveness of both CO_2 capture systems with the reference energy units. The results are shown in Figure 3. The graph describes the rising trend of the price of CO_2 allowance throughout the years 2015–2050. The analyzed CCTs have the potential to achieve cost-effectivity in comparison to the power plant without CCT in the observed time frame. The PCC-A would be cost-competitive in the case the carbon price would be lower than €22.5/tCO_2 (for CCU case), or €26.3/tCO_2 (for CCS case). The IGCC-CaL would be cost-effective in the condition that the carbon price increases up to €58.7/tCO_2 (for CCU case) and €60.9/tCO_2 (for CCS case). PCC-A could achieve cost-effectiveness with carbon price at €24.1/tCO_2 or €20/tCO_2 under the condition of total CAPEX reduction by 5% and 15%, respectively.

It is important to note that the capital investments are decreasing over time for CCT, and therefore, total CCT costs will gradually decrease (dashed green and red lines in Figure 5). These economic dynamics might be the subject of further comprehensive economic study on CCT.

It is assumed that the price of the CO_2 allowance will continue to rise (dashed black line in Figure 3) in the current state of climate crisis and economy crisis. Therefore, the economic decision to invest into the CCT may be major but only in the first years (3–4 years of the CCT construction) but it will lead to cost savings after the payback period (six years according to Reference [4]), as opposed to dealing with the inflating price of CO_2 allowance.

5.3. Environmental and Economic Combination Score

The decision-making process of the CCT integration into power units must be based on a complex assessment, where the environmental and economic scores combine. This combination can be done by simple multiplying the environmental score (sum of all normalized values) and values of OPEX or CAPEX. PCC-A in comparison with IGCC-CaL, has a worse environmental score (0.004) and lower OPEX (€123.06 million) and CAPEX values (€1.097 billion). The IGCC-CaL system has lower values for environmental score (0.0023) but higher values for OPEX (€140.3 million) and CAPEX (€1.2641 billion).

This decision conflict between environmental and economic performance can be resolved by the total product (multiplying the environmental score and CAPEX (or OPEX) value (Figure 4). If the total value is low, it leads to the conclusion that the combination of environmental and economic performance is more favorable for the chosen technology.

The product of environmental score by CAPEX and OPEX can be seen in Figure 6. The graph shows that the IGCC-CaL unit has a smaller total score for both CAPEX and OPEX combination with the environmental score. It concludes that even if the CAPEX and OPEX of IGCC are higher than in the PCC-A process, the environmental performance seems to lower the total combination score.

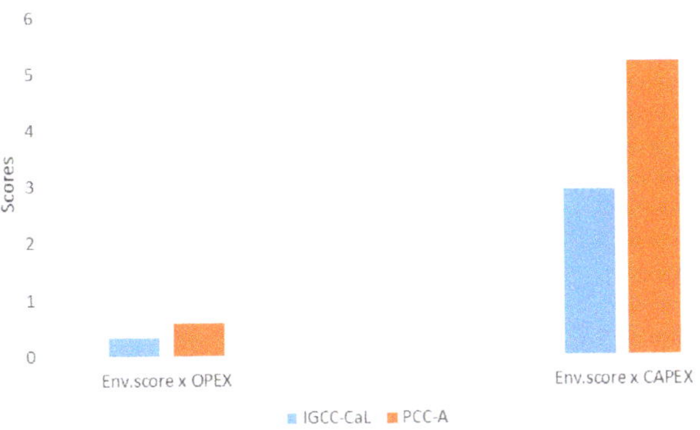

Figure 6. Product of environmental score by CAPEX and OPEX for pre-combustion CO_2 capture integrated into the gasification combined cycle (IGCC-CaL) and post-combustion capture by adsorption of CO_2 by activated carbon (PCC-A).

6. Conclusions

The study presents one of the possible options to select the suitable form of carbon capture technology (CCT) that would meet the sustainability indicators (technical, environmental, and economic) in the Czech energy industry. The presented paper combines environmental and economic variables to conclude the viable choice of carbon capture integration into the brown coal power units. The comparison of both systems concludes into the following points:

- The specific case of IGCC with CaO looping integration has an overall better environmental performance with higher CO_2 ratio capture of 95%; IGCC-CaL decreases the power unit thermal efficiency by 12.5%; IGCC-CaL has a higher capture cost and would become competitive if the carbon price increases up to €58.7/tCO_2 (for CCU case) and €60.9/tCO_2 (for CCS case).
- The specific PCC-A case has an overall worse environmental performance with 75% of CO_2 capture; PCC-A decreases the power unit thermal efficiency by 4.67%; Adsorption has lower capture costs and would become competitive if the carbon price is lower than €22.5/tCO_2 (for CCU case) or €26.3/tCO_2 (for CCS case).
- The cost of electricity of both assessed carbon capture systems is exceeding the current market price.
- The product of the environmental score by CAPEX, and the product of the environmental score by OPEX is lower for IGCC system with CaO looping.

Although it seems that IGCC-CaL integration would be the most suitable option for the carbon capture, thermal efficiency decrease can cause the additional increase in CAPEX and OPEX, which can also result in some environmental burden. On the other hand, the adsorption process can be optimized to enhance the environmental performance with relatively low investments.

It must be stressed that the decision-making process of CCT integration is affected by many other factors such as political decisions, social acceptance, or the economic statements of the energy industry and companies of the Czech Republic. This study presents how an environmental tool such as LCA and economical computation of cost effectiveness may help to contribute to the extensive feasibility study and complex decision-making process.

Further research of CCT technologies integrated into the Czech energy conditions could enhance this research and consider robust techno-economic and environmental analysis of all three considered technologies: post-combustion ammonia scrubbing, post-combustion capture by activated carbon adsorption, and pre-combustion IGCC-CaL.

Author Contributions: Conceptualization, K.Z. and A.C.; methodology, K.Z., A.C., J.Š., M.V., and V.K.; formal analysis, K.Z.; resources, K.Z. and M.V.; data curation, K.Z., J.Š., L.P., and M.V.; writing—original draft preparation, K.Z. and M.V.; writing—review and editing, K.Z., A.C., J.Š., M.V., and L.P.; supervision, V.K. and A.C. All authors have read and agreed to the published version of the manuscript.

Funding: This research received no external funding.

Acknowledgments: This work was supported by the Technology Agency of the Czech Republic, project number TH03020169 and project number TA02020205, and by institutional support from the University of Chemistry and Technology Prague.

Conflicts of Interest: The authors declare no conflict of interest.

Abbreviations

ALO	Agricultural land occupation
AvCo	Avoided cost
CAPEX	Capital expenditures
CC	Climate change
CCo	Capture cost
CCS	Carbon capture and storage
CCT	Carbon capture technology
COE	Cost of electricity

EP	Eutrophication potential
FD	Fossils depletion
FE	Freshwater eutrophication
FET	Fresh water ecotoxicity
FU	Functional unit
GWP	Global warming potential
HT	Human toxicity
IGCC-CaL	Integrated gasification combined cycle with calcium looping
IGCC-FeL	Iron-based oxygen carriers
IR	Ionizing radiation
LCA	Life cycle assessment
MD	Metal depletion
ME	Marine eutrophication
MET	Marine ecotoxicity
NLT	Natural land transformation
OD	Ozone depletion
OPEX	Operational expenditures
PCC-A	Post combustion capture by adsorption
PMF	Particulate matter formation
POF	Photochemical oxidant formation
REU	Reference energy unit
Sc.	Scenario
TA	Terrestrial acidification
TET	Terrestrial ecotoxicity
ULO	Urban land occupation
WD	Water depletion
WTA	Waste heat utilization

References

1. European Association for Coal and Lignite, 2020: The Voice of Coal in Europe—Czech Republic. Available online: https://euracoal.eu/info/country-profiles/czech-republic/ (accessed on 10 June 2020).
2. Carvalho, A.; Mimoso, A.F.; Mendes, A.N.; Matos, H.A. From a literature review to a framework for environmental process impact assessment index. *J. Clean. Prod.* **2014**, *64*, 36–62. [CrossRef]
3. Štefanica, J.; Smutná, J.; Kočí, V.; Macháč, P.; Pilař, L. Environmental gains and impacts of a CCS technology—Case study of post-combustion CO_2 separation by ammonia absorption. *Energy Procedia* **2016**, *86*, 215–218. [CrossRef]
4. Zakuciová, K.; Štefanica, J.; Carvalho, A.; Kočí, V. Environmental Assessment of a Coal Power Plant with Carbon Dioxide Capture System Based on the Activated Carbon Adsorption Process: A Case Study of the Czech Republic. *Energies* **2020**, *13*, 2251. [CrossRef]
5. Singh, B.; Strømman, A.H.; Hertwich, E.G. Comparative life cycle environmental assessment of CCS technologies. *Int. J. Greenh. Gas Control* **2011**, *5*, 911–921. [CrossRef]
6. Cormos, C.C. Integrated assessment of IGCC power generation technology with carbon capture and storage (CCS). *Energy* **2012**, *42*, 434–445. [CrossRef]
7. Falcke, T.J.; Hoadley, A.F.A.; Brennan, D.J.; Sinclair, S.E. The sustainability of clean coal technology: IGCC with/without CCS. *Process Saf. Environ. Prot.* **2011**, *89*, 41–52. [CrossRef]
8. Petrescu, L.; Cormos, C.C. Environmental assessment of IGCC power plants with pre-combustion CO_2 capture by chemical & calcium looping methods. *J. Clean. Prod.* **2017**, *158*, 233–244.
9. Rolfe, A.; Huang, Y.; Haaf, M.; Rezvani, S.; Dave, A.; Hewitt, N.J. Techno-economic and Environmental Analysis of Calcium Carbonate Looping for CO_2 Capture from a Pulverised Coal-Fired Power Plant. *Energy Procedia* **2017**, *142*, 3447–3453. [CrossRef]
10. Clarens, F.; Espí, J.J.; Giraldi, M.R.; Rovira, M.; Vega, L.F. Life cycle assessment of CaO looping versus amine-based absorption for capturing CO_2 in a subcritical coal power plant. *Int. J. Greenh. Gas Control* **2016**, *46*, 18–27. [CrossRef]

11. International Organization for Standardization (ISO). *International Standard ISO 14044 Environmental Management—Life Cycle Assessment—Requirements and Guidelines*; ISO: Geneva, Switzerland, 2006.

12. Pilař, L.; Vlček, Z. *Projektový Návrh Technického Řešení Nízkoemisního Energetického Systému*; ÚJV Řež, a. s.: Husinec, Czech Republic, 2018.

13. Vávrová, J.; Štefanica, J.; Hájek, P.; Smejkalová, J. *Technical Report of Project TA02020205*; ÚJV Řež, a. s.: Husinec, Czech Republic, 2013.

14. Michael, Z.; Hauschild, R.K.; Rosenbaum, S.I.O. *Life Cycle Assessment, Theory and Practice*; Springer International Publishing AG: Montpellier, France, 2018.

15. Tobergte, D.R.; Curtis, S. ILCD Handbook. *J. Chem. Inf. Model.* **2013**, *53*, 1689–1699.

16. Goedkoop, M.; Heijungs, R.; Huijbregts, M.; De Schryver, A.; Struijs, J.; Zelm, R. Van ReCiPe 2008. *Potentials* **2009**, *1*, 1–126.

17. Mimoso, A.F.; Carvalho, A.; Mendes, A.N.; Matos, H.A. Roadmap for Environmental Impact Retrofit in chemical processes through the application of Life Cycle Assessment methods. *J. Clean. Prod.* **2015**, *90*, 128–141. [CrossRef]

18. Vivarová, M.; Smutná, J. *Ekonomické Aspekty Nasazení Systému Záchytu CO_2 na Bázi Karbonátové smy č ky v Elektrárně IGCC s Inovativním Vysokoteplotním čištěním Syntézního Plynu Výzkumná Zpráva*; ÚJV Řež, a. s.: Husinec, Czech Republic, 2017.

19. Vitvarová, M.; Dlouhý, T.; Havlík, J. *TECHNICKO- Ekonomické Posouzení Nasazení Systému Záchytu CO_2 v Elektrárně Výzkumná Zpráva*; ÚJV Řež, a. s.: Husinec, Czech Republic, 2015.

20. Wall, T.; Liu, Y.; Spero, C.; Elliott, L.; Khare, S.; Rathnam, R.; Zeenathal, F.; Moghtaderi, B.; Buhre, B.; Sheng, C.; et al. An overview on oxyfuel coal combustion-State of the art research and technology development. *Chem. Eng. Res. Des.* **2009**, *87*, 1003–1016. [CrossRef]

21. Bradáčová, K.; Macha, P. *Ústav Plynárenství, Koksochemie a Ochrany Ovzduší, VŠCHT Praha, Vysokoteplotní Čištění Energetického Plynu–Semestra Project Souhrn*; University of Chemistry and Technology: Praha, Czech Republic, 2007.

22. Abad, A.; de las Obras-Loscertales, M.; García-Labiano, F.; de Diego, L.F.; Gayán, P.; Adánez, J. In situ gasification Chemical-Looping Combustion of coal using limestone as oxygen carrier precursor and sulphur sorbent. *Chem. Eng. J.* **2017**, *310*, 226–239. [CrossRef]

23. Schjolset, S. The MSR: Impact on Market Balance and Prices Price Forecast: Key Assumptions. 2014. Available online: https://ec.europa.eu/clima/sites/clima/files/docs/0094/thomson_reuters_point_carbon_en.pdf (accessed on 12 June 2020).

24. Stock Market Markets Insider, 2020-CO_2 European Emission Allowances. Available online: https://markets.businessinsider.com/commodities/co2-european-emission-allowances (accessed on 12 June 2020).

25. Rabiega, W.; Sikora, P.; Gąska, J. *CO_2 Emissions Reduction Potential in Transport Sector in Poland and the Eu Until 2050*; The National Centre for Emissions Management (KOBiZE): Warsaw, Poland, 2019.

26. European Environment Agency. *Trends and Projections in the EU ETS in 2018*; Publications Office of the European Union: Luxembourg, 2018.

27. Vesely, L.; Skaugen, G.; Vitvarova, M.; Roussanaly, S.; Novotny, V. Case study of transport options for CO_2 from IGCC coal power plant in the Czech Republic for storage. In Proceedings of the 4th Intenrational conference of chemistry and technology, Mikulov, Czech Republic, 25–26 April 2016; pp. 315–320.

28. Roussanaly, S.; Skaugen, G.; Aasen, A.; Jakobsen, J.; Vesely, L. Techno-economic evaluation of CO_2 transport from a lignite-fired IGCC plant in the Czech Republic. *Int. J. Greenh. Gas Control* **2017**, *65*, 235–250. [CrossRef]

Article

Evaluation of the Environmental Sustainability of a Stirling Cycle-Based Heat Pump Using LCA

Umara Khan [1], Ron Zevenhoven [1,*] and Tor-Martin Tveit [2]

[1]   Process and Systems Engineering, Åbo Akademi University, FI-20500 Turku, Finland; umara.khan@abo.fi
[2]   Olvondo Technology, NO-3080 Holmestrand, Norway; tmt@olvondotech.no
*   Correspondence: ron.zevenhoven@abo.fi; Tel.: +358-2-2153223

Received: 23 June 2020; Accepted: 20 August 2020; Published: 31 August 2020

Abstract: Heat pumps are increasingly seen as efficient and cost-effective heating systems also in industrial applications. They can drastically reduce the carbon footprint of heating by utilizing waste heat and renewable electricity. Recent research on Stirling cycle-based very high temperature heat pumps is motivated by their promising role in addressing global environmental and energy-related challenges. Evaluating the environmental footprint of a heat pump is not easy, and the impacts of Stirling cycle-based heat pumps, with a relatively high temperature lift have received little attention. In this work, the environmental footprint of a Stirling cycle-based very high temperature heat pump is evaluated using a "cradle to grave" LCA approach. The results for 15 years of use (including manufacturing phase, operation phase, and decommissioning) of a 500-kW heat output rate system are compared with those of natural gas- and oil-fired boilers. It is found that, for the Stirling cycle-based HP, the global warming potential after of 15 years of use is nearly −5000 kg CO_2 equivalent. The Stirling cycle-based HP offers an environmental impact reduction of at least 10% up to over 40% in the categories climate change, photochemical ozone formation, and ozone depletion when compared to gas- and oil-fired boilers, respectively.

Keywords: stirling cycle-based heat pump; gas/oil-fired boilers; life cycle assessment; SimaPro; eco-indicator 99

---

1. Introduction

Energy is one of the sectors that pollute and harm the environment the most [1]. A key challenge is to address global environmental problems by supporting energy and environmental conditions in parallel. For sustainable development, it is important to change practices and technologies. Total energy use and efficiency are significant motivating factors for assessing the environmental effect of energy use on the environment. Therefore, it is essential to follow the principles of sustainable development strictly [2]. Aiming at a cleaner and better future, the negative impacts of energy use on the environment can be minimized by implementing the usage of renewable energy sources or by adopting environmentally friendly technologies.

Electricity and heat generation contribute to almost half of the global annual CO_2 emissions [2]. One of the major contributors to climate change is emissions of CO_2 from the energy sector, which were a major topic for discussions at the 21st Conference of Parties (COP21) in Paris from 30 November to 11 December 2015. As an outcome of this conference, initiatives have been taken to address these issues of annual emissions so that global temperature rise will be below 2 °C, and preferably below 1.5 °C. This target can be achieved by substituting fossil fuels, specifically coal, oil, and natural gas, with renewable energy sources [3].

Heat pumps offer an energy-efficient solution to heating and air conditioning as they can use renewable electricity and low value heat that can often be taken freely from surroundings. Since heat pumps rely on transmission of heat rather than generation of heat, they do so at one-quarter of the operational cost of conventional heating or cooling technologies, depending on the efficiency of the heat pump, which is often given as the coefficient of performance (COP).

With the electricity markets introducing more and cheaper electricity from renewable sources, heat pumps are gaining market share, replacing traditional fuel-based heating. This is currently happening in the industry as well, with higher temperatures—e.g., above 120 °C—being a challenge. This is where high temperature heat pumps [4] become of interest. The Stirling-cycle based heat pump has already been shown to be efficient at high temperatures and high temperature lifts (see, for instance, the previous work by the authors [5]).

The manufacturing phase of a heat pump is a key contribution phase for determining the environmental impacts arising throughout a product life cycle [6,7]. However, the importance of the manufacturing phase in a life cycle assessment (LCA) is dependent on the heat pump, its capacity, main components, and efficiency [8]. The environmental impacts of the operational phase are sometimes less than the impacts caused by the production and assembling phase. For cases such as these, an LCA study is greatly suggested to recognize and quantify the environmental impact hotspots along the complete life cycle of process units or products.

According to Linke et al. [9,10], to make improvements in manufacturing processes and to attain environmental benefits, companies should add Environmental Impact Assessment (EIA) to their manufacturing phase of products. EIA is a necessary step for the planning of any technical structure to gain clear insight into the likely environmental impact of the structure. EIA techniques are designed to minimize or avoid the adverse effect of development, a process, or a product on the environment.

A comprehensive study was conducted by Stamford et al. [11] to investigate the life cycle environmental and economic sustainability of Stirling engine micro-CHP (combined heat and power) systems and compare it with conventional energy provision from natural gas boiler and grid electricity. Another study addressed the environmental impacts of domestic Stirling engine micro-CHP integrated with solar photovoltaics and battery storage. They concluded that relative environmental impacts can be reduced by 35–100% by replacing grid electricity and a gas boiler by such integrated system [12]. Other relevant work includes two studies that estimated the CO_2 reduction achievable by Stirling engine and internal combustion engine-based CHP systems, but they did not follow a life cycle approach [13,14].

In this paper, the environmental footprint of an industrial size Stirling cycle-based heat pump is compared to that of natural gas or oil-fired boilers. Environmental footprint as the name suggests is defined as environmental impacts associated with any entity, process, or product. It considers the resources a person/product/process utilizes and the resulting emissions to land, air, and water. The study was made using the SimaPro software [15] for LCA. The construction phase, 1, 8, or 15 years of use, and the decommissioning and recycling are considered.

LCA is increasingly becoming standard procedure environmental footprint analysis and comparison of processes or products. One novelty of this paper is to apply it to an energy technology that has very recently found application at an industrial scale. Sufficient real-time data have recently been produced to make this study possible. While hardly any work on LCA applied to Stirling cycle-based energy technology has been reported in the open literature, this paper addresses a reversed Stirling cycle-based heat pump.

2. Materials and Methods

The description of the system and a description of the LCA-methodology used in this work are given in the two next sections. The goal and scope, as well as a description of the system boundaries, are given in the third and fourth sections.

2.1. System Description

The heat pump studied for this study is a double-acting alpha configuration Stirling cycle heat pump. A Stirling engine is driven at different temperatures by periodic compression and expansion of a working fluid (for this study, helium gas) such that the net transmission of heat energy results in mechanical work. When operated as a heat pump, the process runs in reverse: work (electricity) is used to yield high-temperature heat from low-temperature heat.

The heat pump is comprised of main components such as the heater, regenerator, cooler, and compression and expansion cylinders arranged in a Franchot configuration. The internal heat exchangers include heating, cooling, and regenerating sections in the same unit. The heat exchanger is constituted of stainless-steel tubes while the regenerator is made of a metallic mesh of stainless steel. The heat pump in this study as shown in Figure 1 is used to recover heat and use this heat along with the electrical input (250 kW) to generate steam at 10 bar at an output of 500 kW [16].

Figure 1. The HighLift HTHP from Olvondo Technology installed in the heat pump room at pharmaceutical company AstraZeneca, Gothenburg, Sweden. The nominal heat output from the heat pump is between 450 and 500 kW.

2.2. LCA Methodology

To evaluate the sustainability of heat pump and compare that to other, more conventional but potentially more polluting heaters, reliable scientific tools that consider the entire lifetime of a product are needed. LCA is an evaluation tool that assesses and quantifies environmental impacts associated with product/process throughout the life cycle of a product known as "cradle to grave" analysis. The manufacturing and use of resources (i.e., materials and energy), as well as emission to the environment (land, air, and water), are calculated for each process. The significance of impact on environmental categories such as climate change, human and ecotoxicity can then be assessed for several so-called impact categories.

For this study, the LCA methodology was applied using the commercial software SimaPro 9.0. *SimaPro* is a software tool that collects, analyzes, and monitors the sustainability performance of any product or process. Two types of data are used to model the functional systems: foreground and background data [17]. The foreground data, which include data about technology, efficiency, and installed capacity, were taken from industrial technical data sheets and literature. Background data, which include information about raw material manufacturing and fuel use for transportation, construction, and decommissioning of functional unit, were obtained from the Ecoinvent (v.3.5) database, which is embodied in the SimaPro software. SimaPro contains many LCI (Life cycle impact) databases, besides the well-known Ecoinvent v3 database, e.g., theAgri-footprint database and the ELCD database. The Ecoinvent database is a data source for studies and assessments based on ISO 14040 and 14044. The Ecoinvent LCI data are utilized to conduct the life cycle assessment, water footprint assessment, life cycle management, carbon footprint assessment, environmental performance monitoring, product design, and eco-design or Environmental Product Declarations (EPD) [17]. The ISO standardization makes LCA scientifically well-supported while databases for it are continuously expanding. Limitations of LCA are the need for reference or comparison data in the databases used, which sometimes need to be added by the user, or, for research purposes, lack of data for a new process or product. For the current study, database as well as real-life heating system data were sufficiently available.

The heating systems studied are: (a) Stirling cycle-based heat pump (SE HP); (b) oil boiler (OB); and (c) natural gas-fired boiler (NGB). These three systems were evaluated for locations in Sweden with identical climatic conditions. The choice of the location was according to the location of the heat pump under study, which is in Gothenburg, Sweden. The results of the analysis might be different depending on the chosen location.

The LCA methodology is comprised of four main stages of analysis: (i) defining goal and scope; (ii) data collection for life cycle inventory (LCI); (iii) identifying the environmental impact of all the inputs and outputs (LCIA); and (iv) interpreting the results. Several assessment methods have developed over time to classify and characterize the environmental performance of a system: Eco-indicator 99, EDIP 2003, CML 2001, IMPACT 2002+, ReCiPe Endpoint, CML 2 baseline 2000, BEES, TRACI 2, EDIP 2, etc. [18]. The methods used for the current analysis are IMPACT 2002+ and Eco-indicator 99. IMPACT 2002+ v Q2.2 [19] combines the midpoint/damage-oriented approach to impact categories such as human toxicity, carcinogenic effects, non-carcinogenic effects, respiratory effects (due to inorganics), etc. The Eco-indicator 99 [20] method specifies the environmental impact in numbers or scores. This score is scaled in such a way that each point signifies the annual environmental load of an average [European] citizen. The impact categories to be investigated for this study are *respiratory effects, climate change, ozone layer depletion,* and *acidification* because they have been found to be the significant ones in this kind of studies.

For each phase during the manufacturing, operation, and decommissioning, inventory lists, including raw materials and fuel acquisition/manufacturing and air/water emissions, were computed and categorized into the impact categories. The data for impact categories were from information provided by Pre Consultants [21]. Through characterization, the environmental impacts were determined for each category. Lastly, it was investigated which practice has the major impact on the environment during manufacturing and decommissioning, respectively, as these phases are independent of the duration of use of a product between these two phases.

2.3. Goal and Scope

The scope of the analysis comprises:

- The quantification of resource use and emissions resulting from the manufacturing, operation, and decommissioning of the 500-kW Stirling cycle-based heat pump (SE HP) and oil and natural gas burners (OB, NGB).
- A quantitative comparison between different heat pumps to provide a more complete picture of the potential benefits of the SE HP over NGB and OB from an environmental footprint viewpoint.
- This analysis is related to Swedish conditions. The functional unit employed for this analysis is a boiler. The lifespan of the boiler is assumed to be 15 years or shorter.

2.4. System Boundary

To quantify the impacts of the analyzed unit, system boundaries must be determined. The system boundaries adopted in this study are shown in Figure 2.

Manufacturing: The first phase across the system boundaries is the manufacturing phase. It includes extraction of raw materials, the production of the parts for the unit, transportation activities, the assembly and packaging of components, and testing analysis at the unit.

Operation: The operation phase includes the transport of the HP or boiler to consumers and the processes that follow during the use phase. The processes entail fuel consumption/acquisition, electricity production and usage during boiler operations, the heating process, the hot water cycle, and combustion emissions. Besides water consumption that may occur, the HP also brings about a consumption (due to losses) of helium working fluid.

Maintenance: The transport of engineers to and from the site for maintenance of unit is included. The installation or replacement of any component needed during maintenance is excluded, as the contribution may be assumed negligible. If some of the main components are replaced, e.g., the main motor, this assumption is no longer valid. However, the heat pump is engineered to ensure long-life operation of the large critical components.

Decommissioning: Lastly, the end of life (EoL) phase is also assessed, considering the heat pump's handling activities after the estimated life span use. In the decommissioning process, disassembly, cleaning, repairing of parts, and final disassembly are considered. In principle, this makes the raw materials available for other use. (In the LCA, this typically results in negative values for environmental impact.)

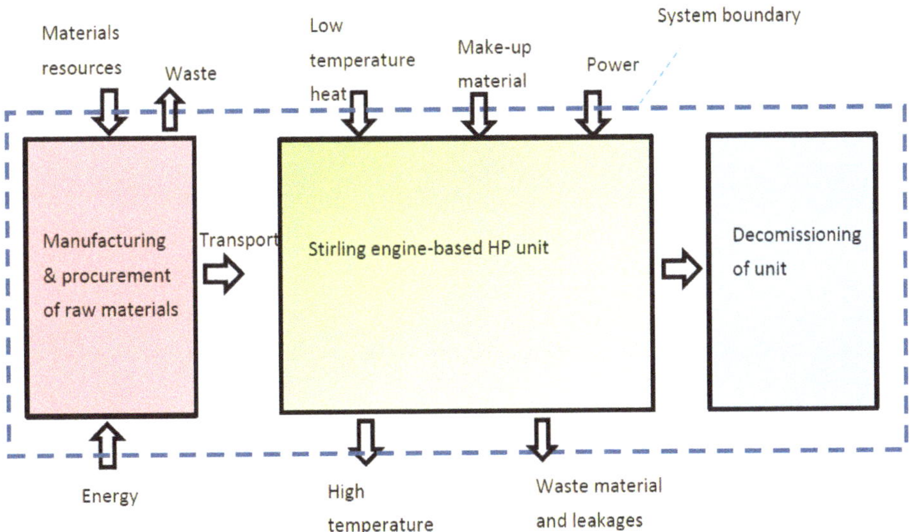

Figure 2. Schematic diagram of the system boundary for life cycle assessment of SE HP.

The system boundary for natural gas-fired boiler and oil boiler are slightly different, consisting of raw material extraction/acquisition, fuel and materials transportation, boiler unit construction, the operation (combustion of fuel), annual maintenance of boiler, and decommissioning of boiler. The production of imported fuel (natural gas) is also considered.

The inventory data sheet used for the development of an LCI network diagram for the Stirling engine-based heat pump (SE HP), natural gas-fired boiler (NGB), and oil boiler (OB) is given in Table 1.

Table 1. Inventory data for LCA analysis of Stirling engine-based HP, Natural gas-fired boiler and oil boiler.

Lifecycle Phase	Raw Materials and Resources Used		SE HP	NGB	OB
Construction	Stainless Steel	kg	7697	104	0
	Steel low-alloyed	kg	0	2396	0
	Steel	kg	0	0	2425
	Cast iron	kg	1500	0	0
	Copper	kg	700	63	125
	Brass	kg	0	1	0.25
	Aluminum	kg	0	156	75
	Lead	kg	0.1	0	0
	Chromium	kg	1	0	0
	Tungsten	kg	1	0	0
	Plastic 1 PTFE	kg	1	0	0
	HDPE	kg	0	19	7
	Silica aerogel	kg	100	0	0
	Rockwool	kg	0	167	95
	alkyd paint	kg	0	0	13
	Brazing solder	kg	0	0	30
	Corrugated board	kg	0	0	50
	Medium voltage electricity	kWh	0	6125	1660
	Natural gas	MJ	0	9833	9600
	Light fuel oil	MJ	0	5187	5100
	Tap water	kg	0	0	3705
Operation	Helium	kg/yr	20	0	0
	Water	m^3/yr	50	0	0
	Motor oil	L/yr	200	0	0
	Land occupation industrial set- up	m^2	50	50	50
	Electricity Power	kW	250	0	0
	Heat output	kW	500	0	0
	Natural gas	MJ/m^3	0	36.8 *	0
	Oil	MJ/kg	0	0	45.2 *

* Based on Ecoinvent v2.0 database (2012).

3. Results

The following section assesses the environmental impacts associated with Stirling cycle-based heat pump, natural gas-fired boiler, and oil boiler with design capacity of 500 kW and compares them with a Stirling cycle-based heat pump. In the tables and figures, dimensionless values are given, grouping several impact categories, unless clearly indicated with a unit.

3.1. Stirling Cycle-Based Heat Pump

The contributions of the construction, operational use, and decommissioning stages of the Stirling cycle-based HP to the total impact were assessed using SimaPro software. Table 2 shows the impacts associated with the generation of 500 kW of heat using a Stirling cycle-based HP.

Table 2. Impact assessment and characterization per impact category for a 500-kW heat output Stirling cycle HP.

Impact Category	Unit	Construction + 1 Year of Use	Construction + 8 Years of Use	Construction + 15 Years of Use + Decommission
Carcinogens	DALY *	2.65×10^{-3}	2.67×10^{-3}	8.80×10^{-4}
Non-carcinogens	DALY	1.01×10^{-2}	1.01×10^{-2}	8.54×10^{-3}
Respiratory (inorganics)	DALY	2.65×10^{-2}	2.69×10^{-2}	1.28×10^{-2}
Ionizing radiation	DALY	6.56×10^{-5}	6.63×10^{-5}	6.17×10^{-5}
Ozone layer depletion	DALY	2.9×10^{-6}	3.20×10^{-6}	2.84×10^{-6}
Respiratory (organics)	DALY	1.80×10^{-5}	2.02×10^{-5}	-6.20×10^{-5}
Aquatic ecotoxicity	PDF $\times$ m^2 $\times$ year **	4.58×10^{2}	4.60×10^{2}	3.60×10^{2}
Terrestrial ecotoxicity	PDF $\times$ m^2 $\times$ year	2.52×10^{4}	2.54×10^{4}	2.17×10^{4}
Terrestrial acid/nutri	PDF $\times$ m^2 $\times$ year	5.42×10^{2}	5.57×10^{2}	3.47×10^{2}
Land occupation	PDF $\times$ m^2 $\times$ year	4.61×10^{2}	4.63×10^{2}	3.22×10^{2}
Global warming	kg CO$_{2\text{-eq}}$	8.11×10^{3}	9.61×10^{3}	-4.89×10^{3}
Non-renewable energy	MJ primary	7.68×10^{5}	7.98×10^{5}	7.00×10^{5}
Mineral extraction	MJ primary	6.82×10^{4}	6.82×10^{4}	6.76×10^{4}

* DALY, disability-adjusted life year; ** PDF, potentially disappeared fraction of species.

Figure 3 shows impact assessment for the heat pump on a relative scale. This means that the plotted values are the values in Table 2 divided by the average for the three cases. As can be seen from the graph, the main contributions to the environmental impact are during the construction and decommissioning stages.

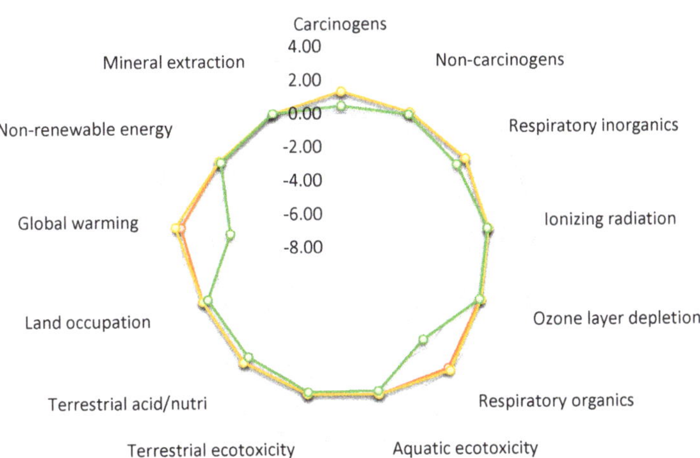

Figure 3. Relative impact assessment for the heat pump with varying use duration. The different factors are made dimensionless by dividing each value by the average value of the factors.

The analysis shows that almost 80% of the impact stems from the production of raw material for constructing the Stirling cycle HP itself, whereas the operation phase contributes less than 20%. Only a small fraction of the impact is due to the maintenance of the engine.

Figure 4 shows that among the impact categories terrestrial ecotoxicity and respiratory inorganic, the share of the processing is close to 50%. The use of water for the operational phase and maintenance phase and the production of cast iron and copper are the main contributing factors, respectively.

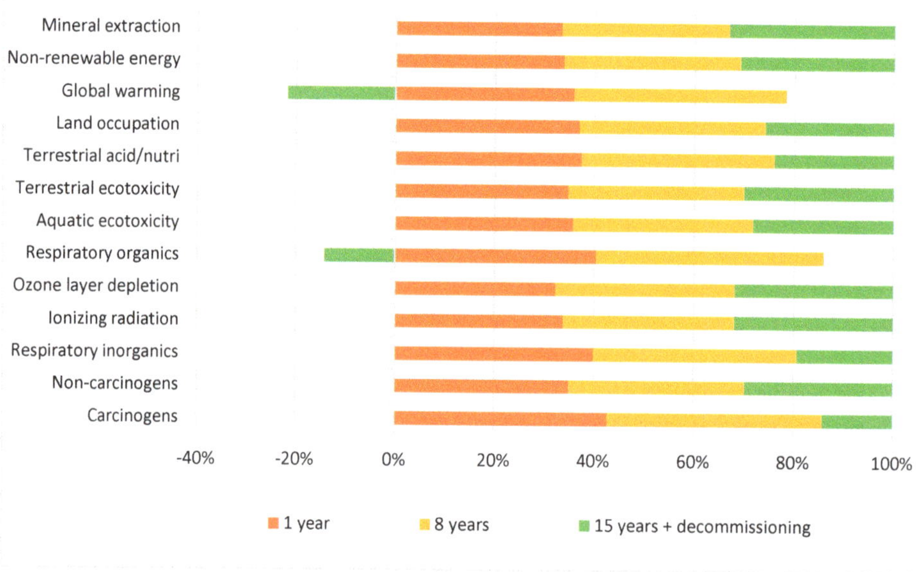

Figure 4. Life-cycle assessment of a 500-kW heat output Stirling cycle-based HP.

When decommissioning is included in the assessment, the impact category global warming and respiratory organic effects show negative values, which means a positive effect on the environment. This effect stems from the 90% recycling of the engine's material.

For most categories, the score is positive, which shows that the net effect is damage to the environment. However, in categories such as respiratory organics and global warming, where a score is negative, the benefits are more significant than the burdens. This is because some substances are paired with a negative characterization factor (C.F.). These substances are known to, for example, contribute to global cooling.

For the Stirling cycle-based HP, the primary emission source leading to the impact is the emissions of zinc to air, mainly stemming from copper production. The analysis showed considerable emission of nitrogen oxides and sulfur oxides as well, which contributes to the photochemical ozone formation and acidification. One of the main contributions to the result is water used for cooling in the context of electricity production.

The resource indium also has a significant impact for the Stirling cycle-based HP. Indium appears in lead-zinc mining as a resource input from nature. In the Ecoinvent dataset, it is assumed that this indium is not used, and thus the resource is wasted. However, with rising demand, it would be possible to extract this resource in the process of lead-zinc mining. The indium accounts for about 60% of the total impact. The contributing factor for ozone depletion by a Stirling cycle-based heat pump is the emission of halons resulting from power generation.

A Stirling cycle-based heat pump has an average impact of 0.02 DALY for human health, 2.2×10^4 PDF·m^2·year. for ecosystem quality, -4894 kg CO$_{2\text{-eq}}$ for global warming and 765,000 MJ for resource consumption. These values include manufacturing, use for 15 years, and decommissioning at end-of-life, as listed in Table 3.

If impacts of the Stirling cycle-based H.P are analyzed over the years, the result shows that one year (including manufacturing phase) of the daily operation of 500-kW heat output Stirling cycle HP emits 8114 kg $CO_{2\text{-eq}}$ with 836,067 MJ energy needed for the extraction/manufacturing of materials. Daily operation of this H.P for eight years (including manufacturing phase) emits 9610 kg $CO_{2\text{-eq}}$ requiring 865,853 MJ energy. Finally, after 15 years of operation including manufacturing and the decommissioning phase, 767,212 MJ energy is needed with overall negative emissions of -4894 kg $CO_{2\text{-eq}}$.

Table 3. The environmental footprint of the 500-kW heat output Stirling cycle HP for construction, use and end-of-life decommissioning.

Damage Category	Unit	Construction + 1 Year of Use	Construction + 8 Years of Use	Construction + 15 Years of Use + Decommission
Human health	DALY *	0.0393	0.0397	0.0223
Ecosystem quality	PDF $\times$ m^2 $\times$ year **	2.66×10^4	2.69×10^4	2.27×10^4
Climate change	kg $CO_{2\text{-eq}}$	0.81×10^4	0.96×10^4	-0.49×10^4

* DALY, disability-adjusted life year; ** PDF, potentially disappeared fraction of species.

The ECO INDICATOR 99 method was used to analyze further the damage on human health, ecosystem quality, and climate change, as shown in Figure 5. The Pt unit (a dimensionless value) measures the impact of these damages. A value of 1 Pt refers to one-thousandth of the yearly environmental impact of one average European inhabitant.

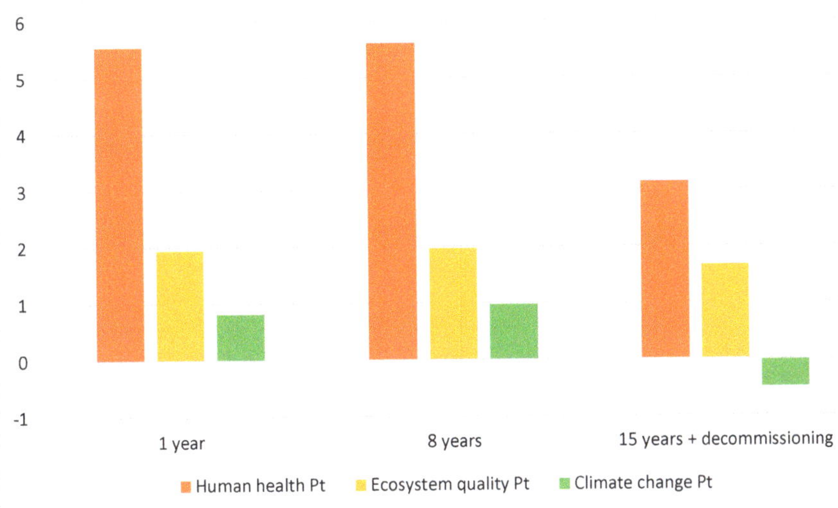

Figure 5. Damage assessment and characterization for Stirling cycle-based HP.

The figure shows that the major impact the Stirling cycle-based HP is on human health. From the analysis of the results, it seems clear that the most critical material in terms of environmental impact is copper (used in the electromotor of the Stirling cycle). The reason is that copper production, although typically 41% recycled copper is used, contributes to the emission of direct atmospheric arsenic emission.

Moreover, the environmental impact for one year of operation is almost the same as for eight years of operation. This shows that the main impact is associated with the production/extraction of raw material for the equipment. It makes clear that, over the 15 years of operation, the additional impact on human health, ecosystem, and climate change is not significant.

3.2. Natural Gas-Fired Boiler (NGB)

A natural gas-fired boiler shows a more significant environmental impact compared to a Stirling cycle HP. For all the options that supply heat by burning natural gas (or oil, as discussed below), the emissions of mercury to air are the crucial values. Table 4 shows the life cycle impacts associated with natural gas boiler for 1, 8, and 15 years of operation. A significant emission source is the emission of bromochlorodifluoromethane. This emission results from the typically long-distance transportation of natural gas in pipelines as it is used for fire suppression within natural gas pipelines infrastructure. The chromium (VI) emissions from iron production process contribute to the human toxicity and cancer effects.

The CO_2 emissions from burning natural gas have the main impact on climate change during boiler use. Some further climate change effects stem from methane emissions that mainly occur due to losses during the transport of natural gas (imported from Denmark via North Sea lines) in long-distance pipelines (methane being the main component of natural gas). The use of a natural gas boiler also results in considerable emissions of particulate matter from the combustion process. Finally, the emission of nitrogen oxides during the combustion process at the heat pump results in photochemical ozone formation, acidification, and terrestrial and marine eutrophication.

Table 4. Impact assessment and characterization for the 500-kW heat output natural gas-fired boiler (NGB).

Damage Category	Unit	Construction + 1 Year of Use	Construction + 8 Years of Use	Construction + 15 Years of Use + Decommission
Human health	DALY *	0.109	0.127	0.114
Ecosystem quality	PDF $\times$ m^2 $\times$ year **	4.08×10^4	4.08×10^4	3.98×10^4
Climate change	kg $CO_{2\text{-eq}}$	8.89×10^5	7.16×10^6	1.27×10^7

* DALY, disability-adjusted life year; ** PDF, potentially disappeared fraction of species.

3.3. Oil Boiler (OB)

The analysis of the life cycle footprint of an oil boiler shows a similar split (Table 5) of the total impact as for natural gas. In addition, here, the emissions of the burning process, especially CO_2, contribute most to the impact category climate change. A prominent difference is that the emissions from the oil burning process also contribute most in the impact categories photochemical ozone formation, terrestrial eutrophication, and marine eutrophication. Electricity (needed during the equipment construction phase) contributes very little in most categories, being also, per MJ of heat produced during the use phase, smaller than for a natural gas boiler. The oil boiler has higher impacts on acidification compared to natural gas, a large extent the result of sulfur dioxide emissions. These emissions result primarily from the oil production (refining) process. For the oil boiler, emissions of copper and zinc to air both contribute to the environmental impact, stemming mainly from the burning process.

For heat from an oil boiler, the emission of bromotrifluoromethane (with a high ozone-depleting potential) from oil production is an important input. The emission stems from leakage, losses at filling, and false alarms.

Table 5. Impact assessment and characterization for oil boiler (OB).

Damage Category	Unit	Construction + 1 Year of Use	Construction + 8 Years of Use	Construction + 15 Years of Use + Decommission
Human health	DALY	0.162	0.21	0.198
Ecosystem quality	PDF $\times$ m^2 $\times$ year	5.45×10^4	5.46×10^4	5.29×10^4
Climate change	kg CO$_{2\text{-eq}}$	1.68×10^6	8.94×10^6	1.73×10^7

The comparison of damage assessment and characterization of Stirling cycle-based HP, oil boiler (OB), and natural gas-fired boiler (NGB) during their life span of 15 years is given in Table 6.

Table 6. Impact assessment and characterization for construction, 15 years of use, and end-of-life decommissioning of a Stirling cycle-based HP (SE HP), an oil boiler (OB), and a natural gas-fired boiler (NGB) for 500-kW heat output.

Impact Category	Unit	NGB	OB	SE HP
Carcinogens	DALY	1.73×10^{-1}	1.59×10^{-1}	8.80×10^{-4}
Non-carcinogens	DALY	1.05×10^{-2}	1.23×10^{-2}	8.54×10^{-3}
Respiratory inorganics	DALY	4.89×10^{-1}	6.75×10^{-1}	1.28×10^{-2}
Ionizing radiation	DALY	2.17×10^{-4}	2.30×10^{-4}	6.17×10^{-5}
Ozone layer depletion	DALY	2.48×10^{-3}	2.76×10^{-3}	2.84×10^{-6}
Respiratory organics	DALY	3.56×10^{-5}	3.60×10^{-5}	-6.20×10^{-6}
Aquatic ecotoxicity	PDF $\times$ m^2 $\times$ year	5.76×10^2	6.66×10^2	3.60×10^2
Terrestrial ecotoxicity	PDF $\times$ m^2 $\times$ year	3.76×10^4	5.01×10^4	2.17×10^4
Terrestrial acid/nutri	PDF $\times$ m^2 $\times$ year	9.99×10^2	1.14×10^3	3.47×10^2
Land occupation	PDF $\times$ m^2 $\times$ year	6.50×10^2	1.04×10^3	3.22×10^2
Global warming	kg CO$_{2\text{-eq}}$	1.27×10^7	1.73×10^7	-4.89×10^3
Non-renewable energy	MJ primary	9.15×10^5	8.12×10^5	7.00×10^5
Mineral extraction	MJ primary	8.14×10^4	1.02×10^5	6.76×10^4

Similar to Figure 3, a comparison of relative impacts (normalized around the average value) for the three technologies is given in Figure 6.

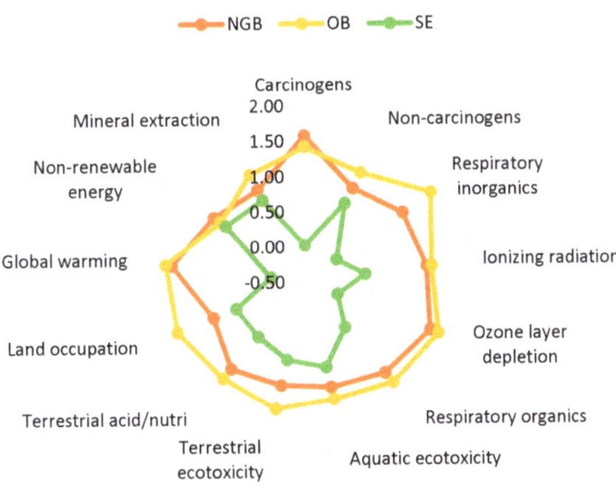

Figure 6. Comparing the relative impact assessment of the technologies. The different factors are made dimensionless by dividing each value by the average value of the factors.

4. Discussion

The Stirling cycle-based heat pump technology causes lower to non-significant environmental impacts compared to a natural gas-fired boiler or an oil-fired boiler. The toxicity originates to the largest part from chromium (VI) emissions into the water in all considered technologies. The unit process is responsible for emissions with the main impact on human toxicity, cancer effects, and global warming.

Figure 7 shows the relative distribution of impacts associated with the use of SE HP, NGB, and OB. Among all categories, human impact is the largest contributing category by these heating technologies following climate change (i.e., global warming potential).

The unit processes with the most significant direct emissions are the processes in which fuels are burned. The largest impacts then come from the oil boiler, followed by the natural gas boiler. The emissions with the highest influence in this category are, besides CO_2, sulfur dioxide and particulate emissions. For the Stirling cycle-based HP construction, nickel and lead manufacturing are the main contributors, besides copper, which is used in the electromotor.

For the impact on climate change, a substantial reduction is possible by replacing a natural gas or oil boiler with a high temperature heat pump. A reduction of up to 15% of the original impact is possible for the options that do not use natural gas. For the oil boiler, a reduction by almost one third is possible. For particulate matter, the oil-fired boiler gives a much higher environmental burden, comparatively. Many impact categories show similar results since the same emissions (see, e.g., emission of nitrogen oxides to air) is responsible for various environmental problems.

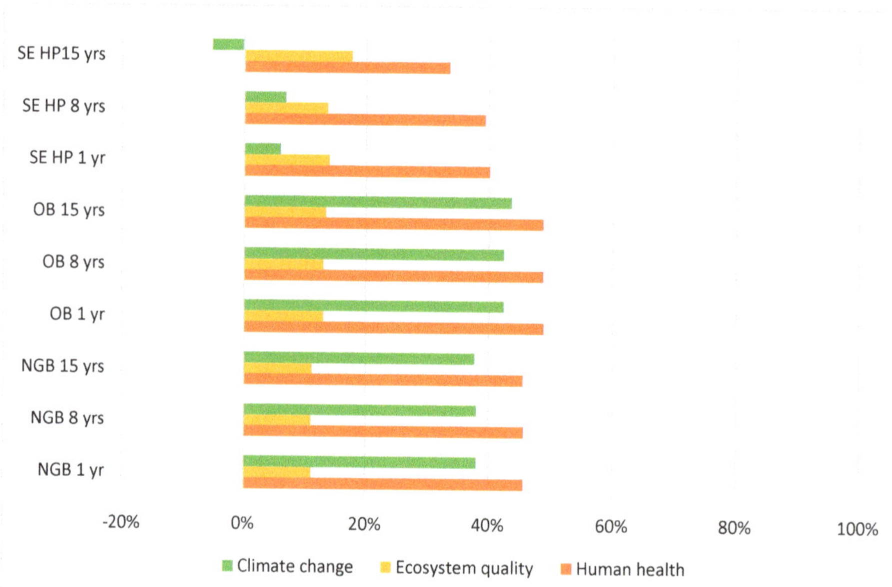

Figure 7. Comparison of damage assessment and characterization Stirling cycle-based HP (SE HP), oil boiler (OB), and natural gas-fired boiler (NGB) on a relative scale.

Figures 8–10 give a comparison of environmental impact for nine damage categories for construction + 1 year of operation (Figure 8), construction + 8 years of operation (Figure 9) and construction + 15 years of operation followed by decommissioning (Figure 10).

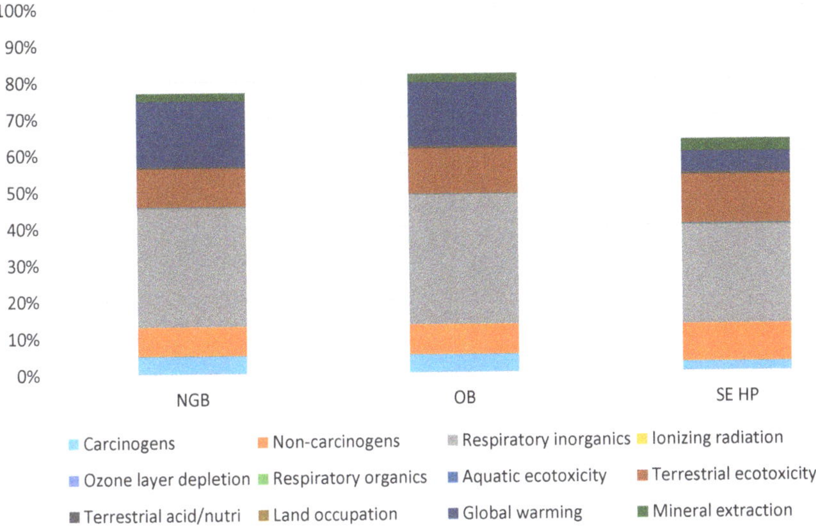

Figure 8. Comparison of damage assessment and characterization for a Stirling cycle-based HP (SE HP), an oil boiler (OB), and a natural gas boiler (NGB) on a relative scale for one year of operation (excluding decommissioning).

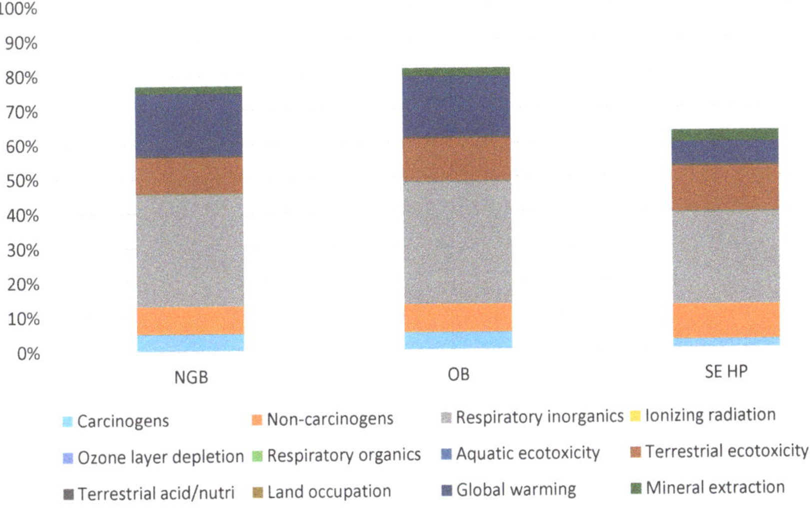

Figure 9. Comparison of damage assessment and characterization Stirling cycle-based HP (SE HP), oil boiler (OB), and natural gas boiler (NGB) on a relative scale for eight years of operation (excluding decommissioning).

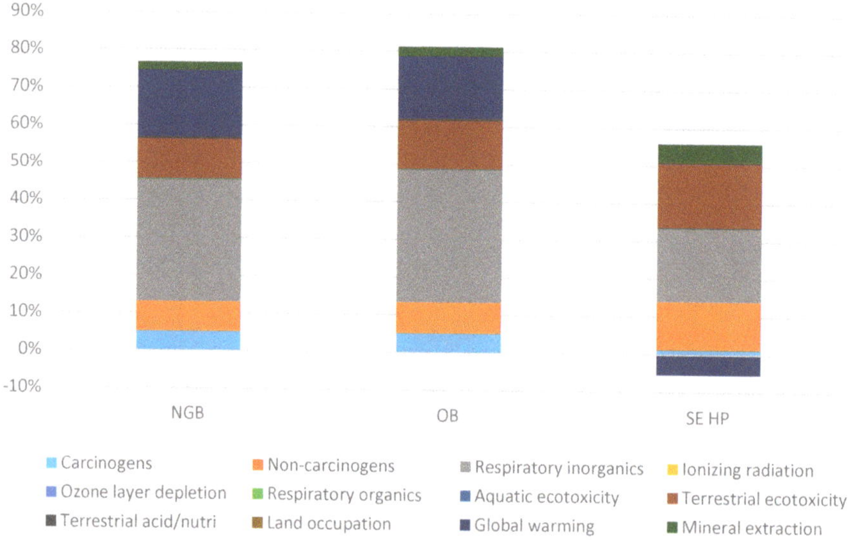

Figure 10. Comparison of damage assessment and characterization for a Stirling cycle-based HP (SE HP), an oil boiler (OB), and a natural gas-fired boiler (NGB) on a relative scale for 15 years of operation (including decommissioning).

The already mentioned paper by Stamford et al. [11] gives a similar study done for a much smaller 1-kW Stirling engine HP and compares it with a gas-fired boiler. They concluded that the S.E. micro-CHP system offers an environmental and economic advantage over the oil boiler by 30%, similar to what is found here.

5. Conclusions

The study evaluated the environmental sustainability of a Stirling cycle-based HP using the LCA approach. The analysis conducted above shows that the manufacturing phase has the most impact during the life span (15 years) of a Stirling cycle-based HP in terms of environmental impacts. The results show that, for the Stirling cycle HP to produce 4 GWh heat output (including manufacturing phase, operation phase, and decommissioning phase), the global warming potential at the end of its life span is −5000 kg CO_2 equivalent and acidification potential 202 kg SO_2 equivalent.

This study also compared the environmental impacts of a Stirling cycle-based heat pump with that of an oil boiler and a natural gas-fired boiler for 500-kW heating. The major impacts of the oil boiler and the natural gas-fired boiler are during the use phase of the engine.

For future work, the comparison should be conducted concerning the economic sustainability of the Stirling cycle-based HP and its comparison with a natural gas-fired boiler, oil boiler, and, if possible, an electric heater boiler. This would be beneficial in providing a still broader picture of how Stirling cycle-based HP technology can replace NGB, OB, and electric boilers in terms of lower environmental and economic impacts.

Author Contributions: Conceptualization, R.Z., U.K. and T.-M.T.; methodology, R.Z. and U.K.; software, U.K., and R.Z.; validation R.Z. and T.-M.T.; formal analysis, U.K. and R.Z.; investigation, U.K., R.Z. and T.-M.T.; resources, U.K., R.Z. and T.-M.T.; data curation, U.K., R.Z. and T.-M.T.; writing—original draft preparation, U.K.; writing—review and editing, R.Z. and T.-M.T.; visualization, T.-M.T.; supervision, R.Z. and T.-M.T.; project administration, R.Z. and T.-M.T.; funding acquisition, R.Z. and T.-M.T. All authors have read and agreed to the published version of the manuscript.

Funding: The European Commission within the Horizon2020 Fast track to innovation project "HIGHLIFT" grant agreement number 831062 funded this work.

Acknowledgments: The authors acknowledge the collaborators Olvondo Technology, AstraZeneca, and Åbo Akademi University for all the necessary support provided to carry out this research work.

Conflicts of Interest: The authors declare no conflict of interest.

Disclosures: One of the authors, T.-M. Tveit, is working in a company developing and marketing the heat pump being studied. All authors are participating in an EU-financed project developing the technology.

References

1. Harbulakova, V.; Zelenakova, M.; Purcz, P.; Olejnik, A. Selection of the best alternative of heating system by environmental impact assessment—Case study. *Environments* **2018**, *5*, 19. [CrossRef]
2. United Nations General Assembly, Transforming our World: The 2030 Agenda for Sustainable Development, Seventieth Session, Agenda Items 15 and 116, 2015. Available online: http://www.un.org/ga/search/view_doc.asp?symbol=A/RES/70/1&Lang=E(pdf) (accessed on 11 May 2020).
3. Günkaya, Z.; Özdemir, A.; Özkan, A.; Banar, M. Environmental performance of electricity generation based on resources: A life cycle assessment case study in Turkey. *Sustainability* **2016**, *8*, 97. [CrossRef]
4. Arpagaus, C.; Bless, F.; Uhlmann, M.; Schiffmann, J.; Bertsch, S. High temperature heat pumps: Market overview, state of the art, research status, refrigerants, and application potentials. *Energy* **2018**, *152*, 985–1010. [CrossRef]
5. Zevenhoven, R.; Khan, U.; Haikarainen, C.; Saeed, L.; Tveit, T.M.; Saxén, H. Performance improvement of an industrial Stirling engine heat pump. In Proceedings of the ECOS 2020—The 33RD International Conference on Efficiency, Cost, Optimization, Simulation and Environmental Impact of Energy Systems, Osaka, Japan, 29 June–3 July 2020; pp. 1042–1053.
6. Diaz, N.; Ninomiya, K.; Noble, J.; Dornfeld, D. environmental impact characterization of milling and implications for potential energy savings in industry. In Proceedings of the 5th CIRP Conference on High Performance Cutting, Zurich, Switzerland, 4–7 June 2012. [CrossRef]
7. Kong, D.; Choi, S.; Yasui, Y.; Pavanaskar, S.; Dornfeld, D.; Wright, P. Software-based tool path evaluation for environmental sustainability. *J. Manuf. Syst.* **2011**, *30*, 241–247. [CrossRef]
8. David, A.D. Moving towards green and sustainable manufacturing. *Int. J. Precis. Eng. Manuf. Green Technol.* **2014**, *1*, 63–66. [CrossRef]
9. Linke, B.; Duscha, M.; Klocke, F.; Dornfeld, D. Combination of speed stroke grinding and high-speed grinding with regard to sustainability. In Proceedings of the 44th CIRP Int. Conference on Manufacturing Systems, Madison, GA, USA, 31 May–3 June 2011.
10. Aurich, J.C.; Linke, B.; Hauschild, M.; Carrella, M.; Kirsch, B. Sustainability of abrasive processes. *CIRP Ann. Manuf. Technol.* **2013**, *62*, 653–672. [CrossRef]
11. Stamford, L.; Greening, B.; Azapagic, A. Life cycle environmental and economic sustainability of stirling engine micro-CHP system. *Energy Technol.* **2018**, *6*, 1119–1138. [CrossRef]
12. Balcombe, P.; Rigby, D.; Azapagic, A. Energy self-sufficiency, grid demand variability and consumer costs: Integrating solar PV, Stirling engine CHP and battery storage. *Appl. Energy* **2015**, *155*, 393–408. [CrossRef]
13. Peacock, A.; Newborough, M. Impact of micro-CHP systems on domestic sector CO_2 emissions. *Appl. Therm. Eng.* **2005**, *25*, 2653–2676. [CrossRef]
14. Peacock, A.; Newborough, M. Effect of heat-saving measures on the CO_2 savings attributable to micro-combined heat and power (μCHP) systems in UK dwellings. *Energy* **2008**, *33*, 601–612. [CrossRef]
15. SimaPro. Available online: https://simapro.com/ (accessed on 4 May 2020).
16. Tveit, T.M.; Khan, U.; Zevenhoven, R. Environmental impact of high temperature industrial heat pumps—From a global warming potential (GWP) perspective, Industrial Efficiency 2020—Accelerating decarbonization. In Proceedings of the ECEEE Industrial Efficiency, Gothenburg, Sweden, 14–16 September 2020.
17. Singh, A.; Miró, L.; Cabeza, L. Environmental Approach. In *High Temperature Thermal Storage Systems Using Phase Change Materials*; Academic Press: Cambridge, MA, USA, 2018; pp. 277–295. [CrossRef]
18. Solé, V.; Ibrahim, D.; Rosen, M. Life cycle assessment of ammonia production methods. *Exergetic Energetic Environ. Dimens.* **2018**, 935–959. [CrossRef]

19. Jolliet, O.; Margni, M.; Charles, R.; Humbert, S.; Payet, J.; Rebitzer, G.; Rosenbaum, R. IMPACT 2002+: A new life cycle impact assessment methodology. *Int. J. Life Cycle Assess.* **2014**, *8*, 324–330. [CrossRef]
20. Eco-Indicator 99 Manual. Available online: https://www.pre-sustainability.com/news/eco-indicator-99-manuals (accessed on 6 May 2020).
21. PRé-Sustainability. Available online: https://www.pre-sustainability.com/ (accessed on 4 May 2020).

Article

Ecological Scarcity Based Impact Assessment for a Decentralised Renewable Energy System

Hendrik Lambrecht *, Steffen Lewerenz, Heidi Hottenroth, Ingela Tietze and Tobias Viere

Institute for Industrial Ecology (INEC), Pforzheim University, Tiefenbronner Str. 65, 75175 Pforzheim, Germany;
steffen.lewerenz@hs-pforzheim.de (S.L.); heidi.hottenroth@hs-pforzheim.de (H.H.);
ingela.tietze@hs-pforzheim.de (I.T.); tobias.viere@hs-pforzheim.de (T.V.)
* Correspondence: Hendrik.lambrecht@hs-pforzheim.de

Received: 30 September 2020; Accepted: 24 October 2020; Published: 29 October 2020

Abstract: Increasing the share of renewable energies in electricity and heat generation is the cornerstone of a climate-friendly energy transition. However, as renewable technologies rely on diverse natural resources, the design of decarbonized energy systems inevitably leads to environmental trade-offs. This paper presents the case study of a comprehensive impact assessment for different future development scenarios of a decentralized renewable energy system in Germany. It applies an adapted ecological scarcity method (ESM) that improves decision-support by ranking the investigated scenarios and revealing their main environmental shortcomings: increased mineral resource use and pollutant emissions due to required technical infrastructure and a substantial increase in land use due to biomass combustion. Concerning the case study, the paper suggests extending the set of considered options, e.g., towards including imported wind energy. More generally, the findings underline the need for a comprehensive environmental assessment of renewable energy systems that integrate electricity supply with heating, cooling, and mobility. On a methodical level, the ESM turns out to be a transparent and well adaptable method to analyze environmental trade-offs from renewable energy supply. It currently suffers from missing quantitative targets that are democratically sufficiently legitimized. At the same time, it can provide a sound basis for an informed discussion on such targets.

Keywords: life cycle impact assessment; distance-to-target weighting; ecological scarcity; renewable electricity and heat generation; decentralized energy system

---

1. Introduction

Diverse criteria need to be considered when planning future energy systems, such as costs, greenhouse gas emissions, land use, and further environmental impacts. As these criteria are measured in different units, they cannot be directly compared and decision-makers need support to consider conflicting targets adequately. This article, therefore, deals with a systematic analysis of environmental trade-offs to better support the design of decentralized renewable energy systems. It is based on a case study conducted as part of the center for applied research "Urban Energy Systems and Resource Efficiency" (ENsource), an inter-university research network that aims to provide scientific support for the design and operation of sustainable energy systems. The case study took place at Mainau GmbH, a tourist company located on an island in Lake Constance in southern Germany. To become climate neutral, its management decided to further increase the share of renewables in the company's energy supply. In order to design an energy system that reduces greenhouse gas (GHG) emissions without increasing other environmental impacts, life cycle assessment (LCA) studies for four future energy supply options were conducted, focusing on the choice of an appropriate metric for evaluating environmental trade-offs.

With the Federal Climate Protection Act [1], the German government has committed itself to become greenhouse gas neutral by 2050. As electricity and heat generation account for more than one-third of German GHG emissions [2], the use of renewable technologies in these sectors is an important lever for the German energy transition. Hence, Mainau GmbH's objectives and strategy are representative of the current efforts of many companies and municipalities in Germany.

Although renewable energies reduce GHG emissions, their construction, disposal, and in some cases, also operation still cause environmental pressures [3,4]. The design of a decarbonized urban energy supply almost inevitably leads to environmental trade-offs. For mineral resources Vidal et al. even point out the danger of a vicious cycle, where "the shift to renewable energy will replace one non-renewable resource (fossil fuel) with another (minerals and metals)" [5]. Hertwich et al. substantiate this concern in an LCA study of a long-term, wide-scale implementation of renewable electricity generation up to 2050 that indicates an increased global consumption of mineral resources like cement, iron, aluminum, and especially copper [6]. As humankind is already at the limits of or even transgressing multiple planetary boundaries [7], a multi-dimensional view of environmental impacts becomes imperative: An energy system design that only takes GHG emissions into account can lead to adverse environmental effects [8].

Hence, methods are required that enable planners and decision-makers to identify energy supply options with a minimal overall environmental impact. LCA provides valuable decision support in this context, as it considers the whole life cycle of power plants and energy carriers [9]. It moreover incorporates a comprehensive environmental impact assessment. Unfortunately, the results of a multi-dimensional life cycle impact assessment (LCIA) are often ambiguous. The main challenge for practical decision support is, therefore, to make impacts in different categories comparable in a meaningful way. Several existing weighting methods can help to solve this dilemma. Even though this approach is generally criticized for necessarily relying on normative value choices [10], it provides valuable decision support in practice [11].

This paper uses the ecological scarcity method (ESM) [12,13] to normalize and weight LCIA results. ESM is a distance-to-target method that weights different environmental pressures based on the ratio of the current situation to the desired policy target. Due to its mathematical simplicity and because the weights depend in a transparent way on publicly available data from laws or environmental authorities [11], it is particularly suitable for communication with practical decision-makers who are usually not LCA experts. For the same reason, it is easier to adapt the ESM to specific decision contexts than other weighting approaches such as, e.g., monetary, panel, or mid-to-endpoint weighting [11,14]. In the present study, we adapt and apply an ESM to renewable energy systems in Germany.

The paper is outlined as follows. Section 2 introduces the main features of ESM and describes essential assumptions and necessary adaptations for the assessment of renewable energy systems in Germany. Sections 3 and 4 deal with the application of the method to the case study of Mainau GmbH. Different energy supply scenarios are ranked regarding their environmental impact and trade-offs between different impact categories are analyzed in more detail. A contribution analysis shows which energy technologies and life cycle processes cause the highest environmental impacts. Based on these results, Section 5 provides recommendations for company management. Section 6 puts the findings from the case study into the broader context of the energy transition, discusses strengths and weaknesses of the adapted ESM, and points out future research needs and fields of application.

2. Materials and Methods

2.1. The Ecological Scarcity Method—Basic Structure and Important Properties

All ESM share the same basic structure [13]: An impact score (IS) is calculated by multiplying elementary flows e_j from a product system's life cycle inventory by specific eco-factors EF_j (Equation (3)) and adding them up (Equation (1)). Elementary flows are material, and energy flows between the system under investigation and the natural environment.

$$IS = \sum_j e_j \cdot EF_j \tag{1}$$

Depending on the set of elementary flows covered by the summation index j, the impact score corresponds either to a specific impact category i (e.g., climate change, mineral resources, water pollution, etc.) or to the total environmental impact. The total impact score IS_{total} thus corresponds to the sum of all category impact scores IS_i (Equation (2))

$$IS_{total} = \sum_i IS_i \tag{2}$$

The eco-factor (EF_j) combines an external normalization with respect to an appropriate reference value (N_j) and a weighting factor: $w_j = \left(A_j/T_j\right)^2$ (Equation (3)). The weight depends on the ratio of the current environmental pressure (A_j) and the desired target value (T_j). ESM thus belongs to the class of distance-to-target weighting methods (cf. [11,15]).

$$EF_j = \cdot \frac{1}{N_j} \cdot \left(\frac{A_j}{T_j}\right)^2 \tag{3}$$

For practical implementation, some specifications and extensions to this basic concept are necessary (Equation (4)): First, most quantities involved in calculating eco-factors refer to a certain region x (e.g., Switzerland, Germany, the World) and time horizon t (e.g., 2010, 2020, or 2050). Second, the index j in Equation (3) refers to single elementary flows, whereas target values, in some cases, only exist on an aggregated level: For instance, greenhouse gas reduction targets apply to different substances and are therefore expressed as global warming potential (CO_2-equivalents). In this case, Equation (4) integrates characterization, i.e., calculating the contribution of elementary flows j to specific impact categories i via the characterization factor CF_{ij} and carries out both normalization and weighting on the impact level. Finally, a region-specific scaling factor $s(x)$ usually assures "reasonable" numerical values for the impact IS.

$$EF_{j,\ x,\ t} = CF_{ij} \cdot \frac{1}{N_{i,x,t}} \cdot \left(\frac{A_{i,x,t}}{T_{i,x,t}}\right)^2 \cdot s_x \tag{4}$$

Equation (4) reveals an important feature of ESM: its adaptability. So far, it has mainly been used to adapt the original Swiss ESM [13] to other countries including Germany [16,17], Thailand [18], China [19], and the EU [20,21].

2.2. Compilation of the ENsource ESM

The following section describes the compilation of an ENsource ESM to evaluate environmental trade-offs in the specific context of decentralized renewable energy systems in Germany at the example of Mainau GmbH. To this end, normalization references (N), current pressures (A), and target values (T) of suitable existing ESM had to be adapted to current German conditions. We chose the ESM developed by Ahbe et al. for Germany as a starting point, whose eco-factors basically required a time update (Equation (4)) [17]. In order to further increase the ESM's coverage of relevant environmental issues, the impact indicators land use, carcinogenic substances into air, heavy metals into air, and ozone layer depletion have been adopted from an ESM developed by Muhl et al. for the European Union [21]. Here, some quantities had to be scaled to German conditions. Eventually, the impact category mineral resources, which is missing in the ESMs of both Ahbe et al. [17] and Muhl et al. [21], was integrated following Frischknecht and Büsser-Knöpfel [13] but applying the latest abiotic resource depletion potentials based on the ultimate reserve according to van Oers et al. [22].

All resulting eco-factors refer to $x =$ Germany. For consistency reasons, the base year is $t = 2017$ for all normalization and current environmental pressures. The time horizon for the target values is $t = 2050$ unless this was not possible due to the lack of data. The attribution of elementary flows to impact categories (classification) follows the impact assessment method "ecological scarcity 2013" as implemented in the ecoinvent v3.5 database [13,23].

Table 1 provides an overview of all considered impact indicators in the ESM. The asterisks in column "Adapted" indicate that most target values (T) have been changed with respect to their respective origins, e.g., if more recent legislative references were available. The column "legitimation" provides a classification of the target value's degree of democratic legitimation according to the following guidelines: A high legitimation (+) is assigned to quantitative targets adopted directly from a national law or a regulation as, e.g., the Federal Climate Protection Act in the case of Global Warming (GW) [1]. Targets based on binding international treaties or guidelines are assigned a medium legitimation (o), e.g., the UN Protocol on Heavy Metals that sets binding targets for Germany for heavy metal emissions into air (HMIA). Eventually, a low legitimation (−) indicates a target derived from a qualitative objective or strategy [24]. This applies, for example, for the targets for mineral resources (MR) and land use (LU), which are deduced respectively from the German resource efficiency program II [25] and sustainability strategy [26]. For a detailed documentation of the eco-factors in the ENsource ESM please refer to the supplementary information (S1_ENsource ESM, S2_ENsouce ESM elementary flows).

Table 1. ENsource ESM impact indicators. Origin: indicates ESM from which indicator was adopted: A = Ahbe et al. [17], M = Muhl et al. [21], F = Frischknecht and Büsser-Knöpfel [13]. Adapted: Asterisks (*) indicate adaptations of target value T with respect to original ESM. Legitimation: Different categories: (+) high (e.g., law, regulation); (o) medium (e.g., binding international treaty, EU directives); (−) low (e.g., derived from qualitative goals, strategies).

Impact Class	Impact Indicator	Abbreviation	Origin	Adapted	Legitimation
Global warming	global warming	GW	A	*	(+)
Ecosystem quality	carcinogenic substances into air	CSIA	M		(−)
	heavy metals into air	HMIA	M	*	(o)
	heavy metals into water	HMIW	A	*	(o)
	main air pollutants and PM	APP	A	*	(o)
	ozone layer depletion	ODP	M		(+)
	water pollutants	WP	A	*	(o)
Resources	mineral resources	MR	F	*	(−)
	water resources	WR	A		(−)
	land use	LU	M	*	(−)
	energy resources	ER	A		(−)
Waste	non-radioactive waste to deposit	WTD	A		(−)

2.3. Comparative Analysis of the ESM Weighting Schemes

Figure 1 compares the normalized weights of the ENsource ESM with its predecessor methods. The selection of impact categories corresponds to the ENsource ESM. Particularly striking are the differences with Ahbe et al., as this was the starting point for the development of the ENsource ESM [17].

First, the significantly higher share of GW in the ENsource ESM weighting scheme is conspicuous. It corresponds to the more ambitious target value for GHG emission reductions (−90% of 1990 GHG emissions by 2050 as compared to −80% in the ESM of Ahbe et al.) based on the following consideration: The Federal Climate Protection Act [1] strives for climate neutrality. To avoid the mathematical singularity and because we assume that a complete implementation is not to be expected, we use the mean value between the target value used by Ahbe et al. and a reduction to zero. Even though this does not correspond exactly to the legal limit, the resulting weight for global warming certainly better reflects the current political priority setting than the original weight in Ahbe et al. [1,17].

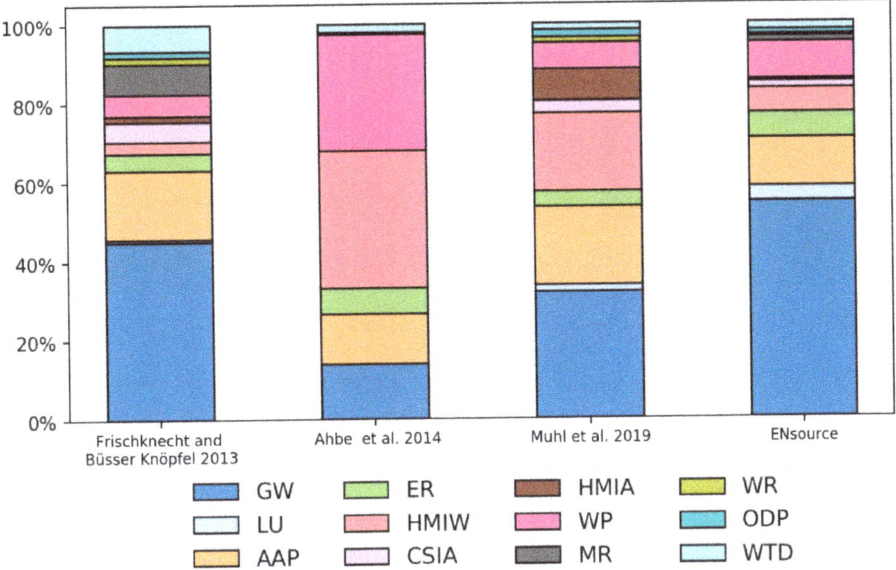

Figure 1. Weighting schemes of the different ecological scarcity methods. GW = global warming, LU = land use, APP = main air pollutants and PM, ER = energy resources, HMIW = heavy metals into water, CSIA = carcinogenic substances into air, HMIA = heavy metals into air, WP = water pollutants, MR = mineral resources, WR = water resources, ODP = ozone layer depletion, WTD = non-radioactive waste to deposit.

Furthermore, the large contribution of water pollutants (WP) and heavy metal emissions into water (HMIW) in the weighting scheme of Ahbe et al. is striking. The corresponding target values stem from a personal communication with the Federal Environment Agency [17]. As they could not be updated with publicly available sources, the ENsource ESM uses the original methodology from Frischknecht and Büsser-Knöpfel instead [13]. The target values for HMIW and polycyclic aromatic hydrocarbons (concerning WP) thus correspond to critical concentrations [27]. The remaining WP target values for phosphorus, nitrogen, and chemical oxygen demand correspond to those in Ahbe et al. [17].

3. Case Study: Renewable Energy Supply Scenarios for Mainau GmbH

In the present study, we analyze four scenarios that result from an energy system optimization carried out by the ENsource partner Fraunhofer Institute for Solar Energy systems (ISE) by means of the optimization tool KomMod [28]. All scenarios imply the coupling of heat and electricity sector. First, an analysis of the renewable energy potential is conducted for the island of Mainau to identify relevant constraints. For instance, due to lack of space for wind turbines on the island, wind energy is excluded from the energy supply portfolio. KomMod then minimizes operational and investment costs under the assumption that no energy infrastructure exists. This greenfield approach shall assure a fair comparison of existing with newly installed energy technologies.

3.1. Energy System Scenario Description

The "Business as usual scenario" (BAU) corresponds to the current energy system of Mainau GmbH. Scenario 1 (S1) integrates a power-to-liquid plant to generate methanol as fuel for the company's vehicle fleet. It uses surplus energy of a photovoltaic plant (PV) installed on the car park roof on the nearby shore. This requirement corresponds to a constraint of an additional 870 MWh of electric energy (Export in Figure 2) in the optimization model. In order to operate the electrolysis continuously, battery storage is used to compensate for the fluctuations in PV electricity generation. Scenario 2 (S2) exclusively uses the PV potential available on the island. Scenario 3 (S3) additionally comprises the PV potential on the car park but—in contrast to S1—without operating the power-to-liquid plant. Table 2 summarizes the underlying potentials and constraints. The electronic supplementary material (S3_Parameter and modeling) provides further details on capacity, generation, and full load hours per technology and scenario.

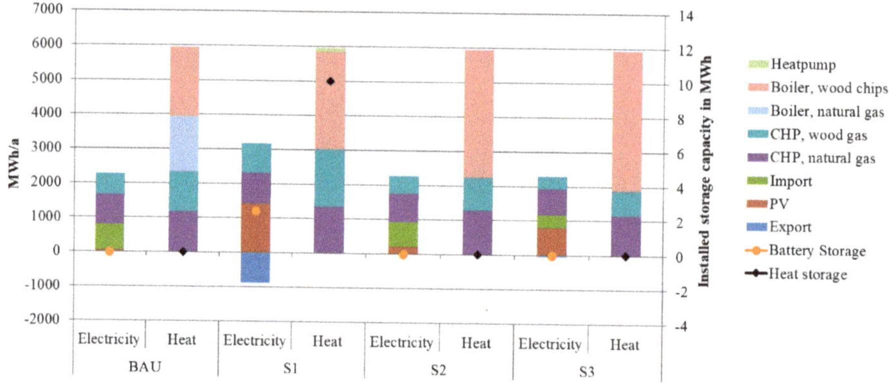

Figure 2. Energy supply and storage technology for all scenarios (CHP = combined heat and power plant; PV = photovoltaic plant).

Table 2. Potentials and restrictions for optimization of the case study scenarios.

Potentials/ Restrictions	Technology	BAU	S1	S2	S3
Potentials	Boiler, natural gas [MWh]	2900	0	0	0
	Boiler, wood chips [MWh]	2000	26,280	26,280	26,280
	Combined heat and power plant, electricity, natural gas [MWh]	880	880	880	880
	Combined heat and power plant, electricity, wood gas [MWh]	600	1314	1314	1314
	Photovoltaic plant [MW]	0.054	3.17	0.23	3.22
	Solarthermal [MW]	0	0.73	0	0
Restrictions	Grid electricity	not restricted	restricted	not restricted	not restricted
	Storages	no	battery, heat	no	no

Figure 2 illustrates the energy technology portfolios for the target year of $t = 2050$ for all scenarios that result from cost optimization. Additionally, Figure 3 provides an overview over the system configurations by means of Sankey diagrams for the energy flows.

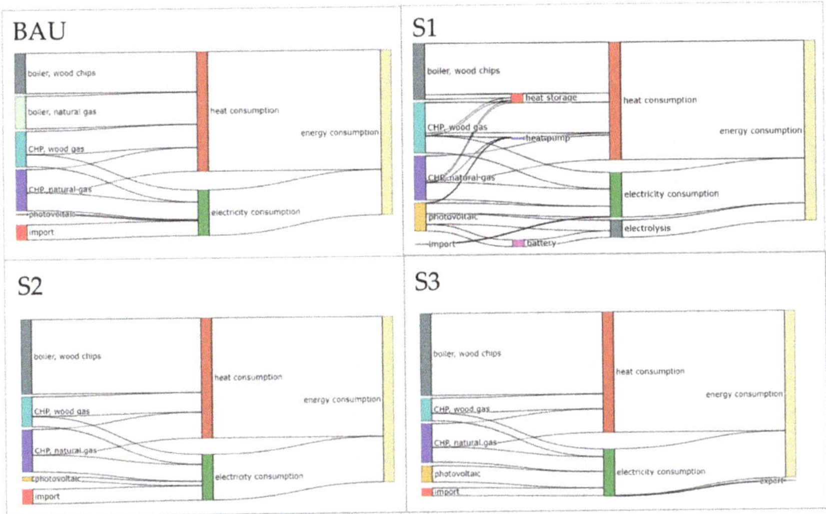

Figure 3. Sankey diagrams of the system configuration.

3.2. Life Cycle Modelling of the Energy System Scenarios

The functional unit to compare the different scenarios is the provision of 2275 MWh of electricity and 5960 MWh of heat (space heating and hot water) per year. It corresponds to the actual energy consumption in 2015, which the Mainau GmbH deems representative for the future. The system boundary is cradle to grave. The inventory analysis uses the software tools Umberto and openLCA, both relying on the database ecoinvent (v3.5, cut-off system model) for background system modeling [23]. In order to assure consistency, process parameters were, wherever possible, set to the same values as in the optimization model (cf. Table 3: Parameterized processes). For PV plants the site-specific yield is taken into account. The combined heat and power (CHP) plant uses a synthetic gas produced from a wood mix harvested in the surroundings of the island. The wood chips boiler uses the same local wood mix. Wood chips production is modeled using generic datasets. Since no wood gas-fired CHP is available in ecoinvent v3.5 the biogas CHP (adapted with site-specific emission data) is combined with a gasifier dataset (see SI_3_Parameter and modeling for modeling details). The allocation of impacts for heat and electricity follows the ecoinvent approach based on exergy. The battery lifetime is set to 10 years [29]. The energy density corresponds to a common lithium battery [30]. Import electricity is represented by the process "market for electricity, low voltage (DE)".

Table 3. Parameterized processes.

Technology	Ecoinvent 3.5 Process	Parameter		
		COP	Energy Density	Lifetime
Heat pump	heat production, air-water heat pump 10 kW	3.4	-	20
Heat pump	heat production, borehole heat exchanger, brine-water heat pump 10 kW	4.1	-	20
Battery storage	battery production, Li-ion, rechargeable, prismatic	-	0.114 kWh/kg	10

Because of missing inventory data, the power to liquid plant is not explicitly modeled in S1. Instead, a credit is given to the 870 MWh surplus electric energy in order to gain functional equivalency between S1 and the other scenarios. As the surplus energy exclusively substitutes fossil fuel combustion in the company's car fleet, the credit is derived from the fuel consumption and CO_2 emissions of a medium-sized diesel car. An estimated methanol production efficiency of 50% and the heating value of diesel are assumed for credit calculation.

4. Results and Discussion

4.1. Scenario Comparison Based on the Total Environmental Impact

The application of the ENsource ESM leads to an unambiguous ranking of the energy supply scenarios according to their total environmental impact (Figure 4). The alternative renewable scenarios only slightly reduce the total environmental impact compared to BAU. However, this merely reflects the fact that Mainau GmbH's current energy system already comprises a high share of renewable energies. To obtain a more complete picture, the renewable scenarios (including BAU) are compared with a fictitious carbon scenario (CS), which corresponds to electricity from the German grid and natural gas-based heat generation. Compared to CS, all renewable scenarios have a significantly lower total environmental burden (up to 30% for S1). Furthermore, the relative share of GW in the total environmental impact is significantly smaller in the renewable scenarios (minimum 26% for S1) than in CS (70%). This underlines the necessity of a multi-dimensional environmental impact analysis when comparing the renewable scenarios with each other. Contrary to the partly significant reductions in almost all other impact categories, land use (up to a factor of 20.3 in the case of S1) and mineral resources (up to a factor 3.2 in the case of S1) increase in all renewable scenarios compared to CS.

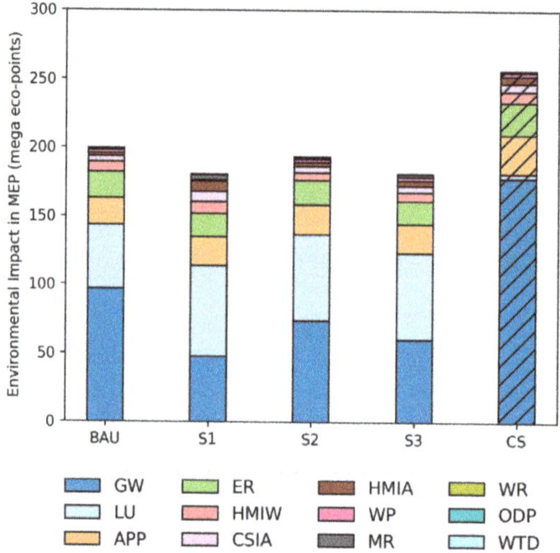

Figure 4. Scenario Impacts with ESM ENsource, GW = global warming, LU = land use, APP = main air pollutants and PM, ER = energy resources, HMIW = heavy metals into water, CSIA = carcinogenic substances into air, HMIA = heavy metals into air, WP = water pollutants, MR = mineral resources, WR = water resources, ODP = ozone layer depletion, WTD = non-radioactive waste to deposit.

The renewable scenarios S1 and S3 have the lowest total environmental impact. The slightly more intensive PV use, the increased usage of the wood chips boiler, and the abandonment of the natural gas boiler in scenario S2 lead to a minor reduction in impacts compared to BAU. Scenario S1 is particularly interesting as it leads to the strongest GHG reductions and is the only scenario that explicitly includes the mobility sector. The following section examines which impact categories particularly contribute to the total environmental impact.

4.2. Comparative Analysis of Environmental Impacts

Based on the ENsource ESM, it is possible to analyze the environmental trade-offs between different scenarios. Figure 4 shows absolute differences between the investigated scenarios compared to BAU for the different impact categories. The global warming (GW) reduction in scenario S3 (−36.2 MEP = mega eco-points) is mainly counteracted by increases in land use (LU, 16.2 MEP), mineral resources (MR) and heavy metal emissions into air and other air pollutants (HMIA and APP, together +4.2 MEP), leaving a clear net environmental benefit (−17.9 MEP). The significantly higher GW reduction in scenario S1 (−49 MEP) is offset by correspondingly higher negative environmental impacts in other categories: land use (LU, +19.0 MEP), mineral resources (MR, +2.7 MEP) and heavy metal emissions (+6.3 MEP both into water and air) but also carcinogenic substances (CSIA, +3.6 MEP). Consequently, despite the clearly better climate performance of S1 its total environmental impact (+18.8 MEP) is only slightly lower than for scenario S3.

To better understand the adverse effects of decarbonizing Mainau GmbH's energy supply and thus to develop context-specific remedies, the following section takes a closer look at critical technologies and life cycle processes that significantly contribute to the increasing impact indicators.

4.3. Hot Spot Analysis: Technologies and Life Cycle Processes

Increasing environmental impacts are mainly due to the additional technical infrastructure required to generate and store renewable energy, which is particularly high for scenario S1 (cf. Figure 5). Precious metals are important drivers of mineral resources (MR): Main consumers are photovoltaic cells (silver for metallization paste), inverters (gold for circuits, copper for converter), the battery (gold for circuits), and the CHP (platinum for catalytic converter). However, the mining and refining processes needed to provide those metals involve other environmental impacts as well, especially heavy metal emissions into air and water (HMIA and HMIW). The battery disposal has significant carcinogenic environmental effects (CSIA). The wood chips boiler and the CHP (both natural and wood gas) contribute mainly to the categories CSIA and other air pollutants (APP). In contrast, increased land use (LU) clearly results from operating the wood chips boiler and, to a lesser extent, the wood gas CHP.

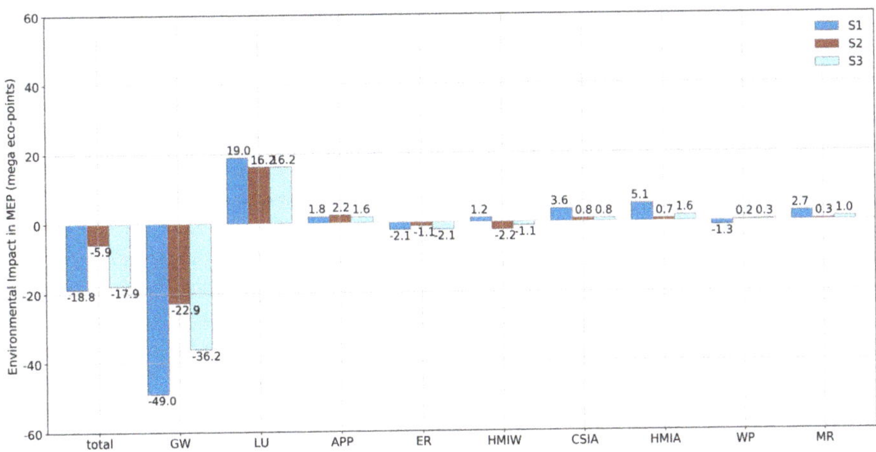

Figure 5. Absolute changes (as differences expressed in MEP = 1e6 EP) of the alternative renewable scenarios compared to BAU, only significant changes considered: GW = global warming, LU = land use, APP = main air pollutants and PM, ER = energy resources, HMIW = heavy metals into water, CSIA = carcinogenic substances into air, HMIA = heavy metals into air, WP = water pollutants, MR = mineral resources.

5. Summary and Recommendations Concerning the Mainau GmbH

The preceding analysis shows that an increased PV use by Mainau GmbH is ecologically beneficial, although GHG savings are partially offset by increased mineral resources and heavy metal emissions (cf. scenario S3 in Figure 6). The inclusion of the mobility sector in scenario S1 makes sense in terms of achieving climate neutrality. However, the continuous operation of the power to liquid system requires a substantial increase of PV capacity, the installation of a heat storage unit, and a battery, as well as the increased use of the wood gasifier. This leads to increases in most of the investigated impact categories that largely offset the GW reductions from substituting fossil fuels. On the other hand, the additional storage capacities entail a completely self-sufficient energy supply. This may be considered an additional advantage in terms of energy systems decentralization, since necessary system services and infrastructure are also partially decentralized. This perspective would put into question the functional unit and system boundaries chosen to compare S1 to the other scenarios.

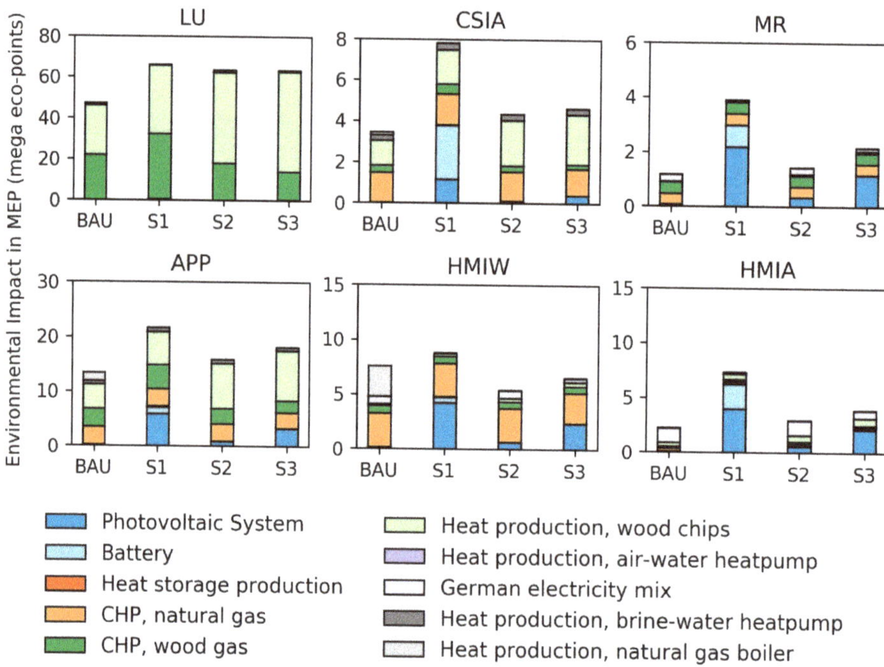

Figure 6. Share of energy carriers and storage technology for several impact categories, LU = land use, CSIA = carcinogenic substances into air, MR = mineral resources, APP = main air pollutants and PM HMIW = heavy metals into water, HMIA = heavy metals into air.

The hot spot analysis reveals the main drivers of negative environmental impacts associated with an increased use of renewable energies: substantial increases in land use (LU), air pollutants (APP, CSIA), and heavy metal emissions (HMIW, HMIA). The latter is closely linked to the mining activities for raw materials for photovoltaic modules and storage technologies. Land use is increasing due to the use of wood as an energy carrier. These are potential levers for further improving the energy system design: e.g., by switching to CdTe cells instead of multi-silicon cells, which cause lower environmental impacts [31] or by installing PV modules from manufacturers that use secondary raw materials or pay attention to high environmental standards in the extraction of raw materials. For the wood gasifier and wood chip boiler, wood waste or wood from extensive forestry use could be used.

The present analysis suggests extending the set of considered scenarios. In particular, the commitment to local energy production seems unnecessarily restrictive. Imported wind energy could be a resource-saving alternative to photovoltaics. The current scenarios are based on an exclusive cost optimization. The integration of the environmental indicators derived from the ENsource ESM as optimization objectives or constraints could help to identify even more environmentally friendly energy supply scenarios.

6. Conclusions and Outlook

6.1. Environmental Trade-Offs of Renewable Energy Systems

This paper presents an updated ESM for Germany, which includes mineral resource use of renewable energy systems. From a practical point of view, the ENsource ESM quantifies environmental trade-offs between different renewable technologies and scenarios and provides a single score for an unambiguous environmental ranking of different options. Eco-factors that integrate external normalization and distance-to-target weighting make different impact categories comparable in terms of their relevance for a specific decision context. Applying the ENsource ESM to the case of a decentralized renewable energy system confirms the increased demand for mineral resources found by other authors [3,6]. However, it also shows that the major contribution to the total environmental impact stems from pollutant emissions associated with the mining activities. In addition, increasing land use due to biomass combustion is emerging as a potential problem area. The shift in importance from mineral resources to pollutant emissions related categories and land use could be due to extending renewable energy system analysis from electricity production to heating and cooling as well as mobility. To explore the findings' generalizability from the case study, the ENsource ESM could be applied to other energy systems in Germany.

6.2. Missing Targets—Limitations and Transparency

As for other ESM [17,21], the presented environmental assessment is mainly limited by missing legally binding, quantitative targets for some impact categories (cf. Table 1). In these cases, provisional targets were derived in the most plausible way possible from suitable references, such as EU regulations or directives or governmental strategy reports. However, this clearly limits the democratic legitimacy, which is one of the most important arguments in favor of ESM in general [32]. This study suggests nevertheless that the ESM's mathematical simplicity makes it a particularly transparent approach that allows for a critical reflection and, if necessary, case-specific adaptation of weights for environmental impacts. It moreover makes it possible to involve practical decision-makers in this process by presenting and communicating key assumptions in a structured way. In that sense, we take in the following a closer look at the weights for heavy metal emissions and land-use, which have been identified in the case study as the main adverse environmental effects of renewable technologies, as well as the weight for global warming, which is without doubt particularly important and involves a legal zero-emission target.

With regard to heavy metal emissions into water, it can be objected that the ENsource ESM does not regionalize weights. This means that policy targets (T) and pollutant emissions (A) for the Rhine are effectively transferred locally to the Mainau and, more questionably, in terms of representativeness, to the entire world. The latter directly affects the validity of our results because pollutant emissions from mining processes occur outside Germany. As environmental standards are lower in many countries than in Germany or even do not exist ($T = \infty$), regionalized weights would be much smaller or even zero. This would improve the environmental score or the investigated renewable scenarios. However, we follow the argumentation of Frischknecht and Büsser-Knöpfel, who prefer in such cases to apply weights derived from German environmental standards in order to prevent "environmental dumping" [13].

The indicator for land use in the ENsource ESM refers to land occupation. Since land occupation is not limited by law in Germany, the corresponding weight in the eco-factor is based on a limit for land transformation (measured in ha/a) taken from Germany's sustainability strategy instead [26]. Other approaches to derive the weighting in a more consistent way would be conceivable: e.g., to relate the worldwide land use by German consumption (A) to the usable area available in Germany (T). Regarding our results, this would tend to increase the contribution of land use to the total impact but leave the ranking of the alternative renewable scenarios (S1, S2, and S3) unchanged. Even though methodically appealing, this approach has been abandoned for the present study, as it lacks democratic legitimacy.

The ESM's transparent way of defining weights is even an advantage where the legal basis is clear. Regarding policies aimed at climate neutrality, the fundamental methodological question arises as to how to deal sensibly with a zero-emission target. Mathematically, this leads to infinite weights (Equation (2)) and, thus, eventually to binary weighting. As pointed out initially, there are good reasons to refuse such a narrowing of the focus. At the same time, the practical way of dealing with this limit within the ENsource ESM provides evidence that a strong emphasis on climate change, does not necessarily lead to neglecting other environmental impacts.

6.3. From Decision to Policy Support

A possible approach to improving the ENsource ESM is the further development of the provisionally derived eco-factors. However, this would contradict the basic idea of ESM, which recommends the use of democratically legitimated target values. Rather than research, a political process is necessary that properly involves science, civil society, and other stakeholders to discuss and finally define such target values for different environmental pressures in a comprehensive and consistent way. As has been pointed out before, the role of science is not to provide normative targets [20]. Nevertheless, science can provide a transparent framework to support an informed discussion on these targets. In that sense, the presented ENsource ESM not only aims at practical decision-support but also at supporting such a political process for developing a consistent set of environmental targets for the necessary transition towards a renewable and sustainable energy supply.

Supplementary Materials: The following are available online at http://www.mdpi.com/1996-1073/13/21/5655/s1, MS Word file S1: ENsource ESM, MS Excel file S2: ENsouce ESM elementary flows, MS Excel file S3: Parameter and modeling.

Author Contributions: Conceptualization and methodology, H.L. and S.L.; life cycle modelling, H.H. and S.L.; writing—original draft preparation, H.L. and S.L.; writing—review and editing, H.L., S.L., H.H., T.V., and I.T.; All authors have read and agreed to the published version of the manuscript.

Funding: This research was funded by the Federal State of Baden-Württemberg—Ministry for Science, Research and Art and European Regional Development Fund—Innovation and Energiewende, in the context of the center for applied research "Urban Energy Systems and Resource Efficiency" (ENsource-2, grant number FEIH_ZAFH_1248932).

Acknowledgments: The authors would like to thank all partners of the ENsource consortium for many fruitful discussions on the issues covered by this article. We particularly acknowledge the energy system scenarios provided by the Fraunhofer Institute for Solar Energy Systems (ISE) based on evaluating the optimization model KomMod.

Conflicts of Interest: The authors declare no conflict of interest. The funders had no role in the design of the study; in the collection, analyses, or interpretation of data; in the writing of the manuscript, or in the decision to publish the results.

References

1. Climate Protection Act. KSG, Bundesgesetzblatt. Available online: https://www.bgbl.de/xaver/bgbl/start. xav?startbk=Bundesanzeiger_BGBl#__bgbl__%2F%2F*%5B%40attr_id%3D%27bgbl119s2513.pdf%27% 5D__1603959900484 (accessed on 26 October 2020).

2. Environment Federal Agency. Reporting under the United Nations Framework Convention on Climate Change and the Kyoto Protocol 2019. *Data on the Environment: Data from German Reporting of Atmospheric Emissions—Air Pollutants (1990–2017)*. 2019. Available online: https://www.umweltbundesamt.de/daten/klima/treibhausgas-emissionen-in-deutschland (accessed on 5 December 2019).

3. Laurent, A.; Espinosa, N.; Hauschild, M.Z. LCA of Energy Systems. In *Life Cycle Assessment: Theory and Practice*; Hauschild, M.Z., Rosenbaum, R.K., Olsen, S.I., Eds.; Springer International Publishing: Cham, Switzerland, 2018; pp. 633–668. ISBN 978-3-319-56475-3.

4. Raugei, M.; Kamran, M.; Hutchinson, A. A Prospective Net Energy and Environmental Life-Cycle Assessment of the UK Electricity Grid. *Energies* **2020**, *13*, 2207. [CrossRef]

5. Vidal, O.; Goffé, B.; Arndt, N. Metals for a low-carbon society. *Nature Geosci* **2013**, *6*, 894–896. [CrossRef]

6. Hertwich, E.G.; Gibon, T.; Bouman, E.A.; Arvesen, A.; Suh, S.; Heath, G.A.; Bergesen, J.D.; Ramirez, A.; Vega, M.I.; Shi, L. Integrated life-cycle assessment of electricity-supply scenarios confirms global environmental benefit of low-carbon technologies. *Proc. Natl. Acad. Sci. USA* **2015**, *112*, 6277–6282. [CrossRef] [PubMed]

7. Rockström, J.; Steffen, W.; Noone, K.; Persson, Å.; Chapin, F.S., III; Lambin, E.; Lenton, T.M.; Scheffer, M.; Folke, C.; Schellnhuber, H.J.; et al. Planetary Boundaries: Exploring the Safe Operating Space for Humanity. *E&S* **2009**, *14*. [CrossRef]

8. Algunaibet, I.M.; Pozo, C.; Galán-Martín, Á.; Huijbregts, M.A.J.; Mac Dowell, N.; Guillén-Gosálbez, G. Powering sustainable development within planetary boundaries. *Energy Environ. Sci.* **2019**, *12*, 1890–1900. [CrossRef] [PubMed]

9. Murphy, D.; Carbajales-Dale, M.; Moeller, D. Comparing Apples to Apples: Why the Net Energy Analysis Community Needs to Adopt the Life-Cycle Analysis Framework. *Energies* **2016**, *9*, 917. [CrossRef]

10. Deutsches Institut für Normung, e.V. *Umweltmanagement –Ökobilanz—Grundsätze und Rahmenbedingungen (ISO 14040:2006) Deutsche und Englische Fassung EN ISO 14040:2006*; Beuth Verlag GmbH: Berlin, Germany, 2019; DIN EN ISO 14040.

11. Pizzol, M.; Laurent, A.; Sala, S.; Weidema, B.; Verones, F.; Koffler, C. Normalisation and weighting in life cycle assessment: Quo vadis? *Int. J. Life Cycle Assess.* **2017**, *22*, 853–866. [CrossRef]

12. Müller-Wenk, R. *Die ökologische Buchhaltung. Ein Informations- und Steuerungsinstrument für Umweltkonforme Unternehmenspolitik*; Campus: Frankfurt, Germany, 1978; ISBN 9783593322506.

13. Frischknecht, R.; Büsser Knöpfel, S. *Swiss Eco-Factors 2013 According to the Ecological Scarcity Method. Methodological Fundamentals and Their Application in Switzerland*; Umwelt-Wissen; Federal Office for the Environment: Bern, Switzerland, 2013. Available online: http://www.bafu.admin.ch/publikationen/publikation/01750/index.html?lang=de (accessed on 11 February 2016).

14. Sala, S.; Cerutti, A.K.; Pant, R. *Development of a Weighting Approach for the Environmental Footprint*; Publications Office of the European Union: Luxembourg, 2018; ISBN 978-92-79-68042-7.

15. Castellani, V.; Benini, L.; Sala, S.; Pant, R. A distance-to-target weighting method for Europe 2020. *Int. J. Life Cycle Assess.* **2016**, *21*, 1159–1169. [CrossRef]

16. Grinberg, M.; Ackermann, R.; Finkbeiner, M. Ecological Scarcity Method: Adaptation and Implementation for Different Countries. *Sci. J. Riga Tech. Univ. Environ. Clim. Technol.* **2012**, *10*, 9–15. [CrossRef]

17. Ahbe, S.; Schebek, L.; Jansky, N.; Wellge, S.; Weihofen, S. *Methode der Ökologischen Knappheit für Deutschland. Umweltbewertungen in Unternehmen; eine Initiative der Volkswagen AG*; 2, überarbeitete Auflage; Logos-Verlag: Berlin, Germany, 2014; ISBN 978-3-8325-3845-3.

18. Lecksiwilai, N.; Gheewala, S.H. A policy-based life cycle impact assessment method for Thailand. *Environ. Sci. Policy* **2019**, *94*, 82–89. [CrossRef]

19. Wang, H.; Hou, P.; Zhang, H.; Weng, D. A Novel Weighting Method in LCIA and its Application in Chinese Policy Context. In *Towards Life Cycle Sustainability Management*; Finkbeiner, M., Ed.; Springer: Dordrecht, The Netherlands, 2011; pp. 65–72. ISBN 978-94-007-1898-2.

20. Ahbe, S.; Weihofen, S.; Wellge, S. *The Ecological Scarcity Method for the European Union. A Volkswagen Research Initiative: Environmental Assessments*; Springer: Berlin/Heidelberg, Germany, 2018; ISBN 973-3-658-19505-2.

21. Muhl, M.; Berger, M.; Finkbeiner, M. Development of Eco-factors for the European Union based on the Ecological Scarcity Method. *Int. J. Life Cycle Assess.* **2019**, *24*, 1701–1714. [CrossRef]

22. van Oers, L.; Guinée, J.B.; Heijungs, R. Abiotic resource depletion potentials (ADPs) for elements revisited—Updating ultimate reserve estimates and introducing time series for production data. *Int. J. Life Cycle Assess.* **2019**. [CrossRef]

23. Wernet, G.; Bauer, C.; Steubing, B.; Reinhard, J.; Moreno-Ruiz, E.; Weidema, B. The ecoinvent database version 3 (part I): Overview and methodology. *Int. J. Life Cycle Assess.* **2016**, *21*, 1218–1230. [CrossRef]

24. UNITED NATIONS. *Protocol on Heavy Metals, as Amended on 13 December 2012. ECE/EB.AIR/115*; UNITED NATIONS: San Francisco, CA, USA, 2014.

25. Federal Ministry for the Environment, Nature Conservation, Building and Reactor Safety. *German Resource Efficiency Program II. Program for Sustainable Use and Protection of Natural Resources*; Federal Ministry for the Environment, Nature Conservation, Building and Reactor Safety: Berlin, Germany, 2016.

26. Federal Government Germany. German Sustainability Strategy 2018. 2018. Available online: https://www.bundesregierung.de/resource/blob/975292/1559082/a9795692a667605f652981aa9b6cab51/deutsche-nachhaltigkeitsstrategie-aktualisierung-2018-download-bpa-data.pdf?download=1 (accessed on 5 December 2019).

27. ICPR. Comparison of the Actual State with the Target State of the Rhine 1990 to 2008 No. 193. 2011. Available online: https://www.iksr.org/fileadmin/user_upload/DKDM/Dokumente/Fachberichte/DE/rp_De_0193.pdf (accessed on 9 December 2019).

28. Eggers, J.-B. *Das kommunale Energiesystemmodell KomMod. Dissertation*; Technische Universität Berlin; Fraunhofer-Institut für Solare Energiesysteme; Fraunhofer IRB-Verlag: Berlin, Germany, 2016.

29. Baumann, M.; Peters, J.F.; Weil, M.; Grunwald, A. CO 2 Footprint and Life-Cycle Costs of Electrochemical Energy Storage for Stationary Grid Applications. *Energy Technol.* **2017**, *5*, 1071–1083. [CrossRef]

30. Notter, D.A.; Gauch, M.; Widmer, R.; Wäger, P.; Stamp, A.; Zah, R.; Althaus, H.-J. Contribution of Li-ion batteries to the environmental impact of electric vehicles. *Environ. Sci. Technol.* **2010**, *44*, 6550–6556. [CrossRef] [PubMed]

31. Wade, A.; Stolz, P.; Frischknecht, R.; Heath, G.; Sinha, P. The Product Environmental Footprint (PEF) of photovoltaic modules-Lessons learned from the environmental footprint pilot phase on the way to a single market for green products in the European Union. *Prog. Photovolt Res. Appl.* **2018**, *26*, 553–564. [CrossRef]

32. Andreas, R.; Sala, S.; Jungbluth, N. Normalization and weighting: The open challenge in LCA. *Int. J. Life Cycle Assess.* **2020**, *25*, 1859–1865. [CrossRef]

Publisher's Note: MDPI stays neutral with regard to jurisdictional claims in published maps and institutional affiliations.

MDPI

St. Alban-Anlage 66

4052 Basel

Switzerland

Tel. +41 61 683 77 34

Fax +41 61 302 89 18

www.mdpi.com

Energies Editorial Office

E-mail: energies@mdpi.com

www.mdpi.com/journal/energies

9 783036 505244